ANTIGENIC VARIATION IN INFECTIOUS DISEASES

Special Publications of the Society for General Microbiology

Publications Officer: Dr. Duncan E.S. Stewart-Tull, Harvest House, 62 London Road, Reading RG1 5AS, UK

Publisher: Academic Press

1. Coryneform Bacteria
2. Adhesion of Micro-organisms to Surfaces
3. Microbial Polysaccharides and Polysaccharases
4. The Aerobic Endospore-forming Bacteria: Classification and Identification
5. Mixed Culture Fermentations
6. Bioactive Microbial Products: Search and Discovery
7. Sediment Microbiology
8. Sourcebook of Experiments for the Teaching of Microbiology
9. Microbial Diseases of Fish
10. Bioactive Microbial Products 2: Development and Production
11. Aspects of Microbial Metabolism and Ecology
12. Vectors in Virus Biology
13. The Virulence of Escherichia coli
14. Microbial Gas Metabolism
15. Computer-Assisted Bacterial Systematics
16. Bacteria in Their Natural Environments
17. Microbes in Extreme Environments
18. Bioactive Microbial Products 3: Downstream Processing

Publisher: IRL Press

19. Antigenic Variation in Infectious Diseases

This book is based on a symposium of the S.G.M. held in Nottingham, September 1985.

SPECIAL PUBLICATIONS OF THE SOCIETY FOR GENERAL MICROBIOLOGY
VOLUME 19

ANTIGENIC VARIATION IN INFECTIOUS DISEASES

Edited by

T.H. Birkbeck

Department of Microbiology, University of Glasgow, Glasgow, UK

and

C.W. Penn

Department of Microbiology, University of Birmingham, Birmingham, UK

1986
Published for the
Society for General Microbiology
by

IRL PRESS
Oxford · Washington DC

IRL Press Limited
P.O. Box 1
Eynsham
Oxford OX8 1JJ
England

British Library Cataloguing in Publication Data

Antigenic variation in infectious diseases.—
(Special Publications of the Society for General Microbiology; 19)
1. Communicable diseases in animals
2. Mammals—Diseases 3. Communicable diseases—Immunological aspects 4. Antigens
I. Birkbeck, T.H. II. Penn, C.W. III. Society for General Microbiology IV.Series
599.02′92 SF781

ISBN 1-85221-000-1 (hardbound)
ISBN 0-947946-99-3 (softbound)

Cover illustration The upper part shows a structural model for GIAV virion based on Parekh *et al.* (1980) and Montelaro *et al.* (1982) RT-reverse transcriptase (see Chaper 4). The lower part shows the major components of the gonococcal outer membrane showing an aqueous pore formed by protein I and conserved and variable regions of protein II (see Chapter 6).

Printed by Information Printing Ltd, Oxford, UK

Preface

Although one or two examples of antigenic variability of micro-organisms in the course of infection have been well-known historically, until recently there has not been a widespread appreciation of the importance of processes of rapid genetic change in a variety of important pathogens. This is surprising in view of the emphasis which has long been placed on the capacity of microbial populations to undergo rapid genetic adaptation to environmental changes, not least due to their high multiplication rates and the large populations involved in many biological processes. The work described in this book, which arose from the Symposium "Antigenic Variation in the Course of Infectious Diseases: a Survival Strategy for Pathogenic Micro-organisms" held in Nottingham in September 1985, amply illustrates that the fascinating phenomena of rapid genetic change and rearrangement are central factors in determining the outcome of many infectious processes.

Historically, the best-known examples of antigenic variation in infection were recognised because clinical observation suggested repeated peaks and troughs in microbial numbers, which could only be explained by emergence of new antigenic types able to evade the immune response. The prime example in this category is the relapsing fever borreliae, now well-investigated at the molecular level and one of the subjects of this volume. Similarly, the elucidation of antigenic variation in African trypanosomes followed careful biological observation of the course of infection, and extensive study of the molecular biology of the process of variation made this probably the best understood example until recently, despite the relative complexities of eucaryotic cell biology. Another example of progress which has stemmed from careful clinical studies over a number of years is the advance in understanding of the molecular basis of the slow virus diseases of domestic animals. Although not directly related to processes of disease, research into the genetic basis of flagellar phase variation in *Salmonella* also commenced early, and the editors regret that it was not possible to include a contribution on this work which has been seminal in our understanding of some of the genetic processes described here.

The past ten years have seen a wave of new developments in research into antigenic variation. The reasons for this are twofold. Firstly, large scale initiatives involving many biologists and institutions of high quality to investigate the pathogenicity and immunobiology of organisms such as the gonococcus have led to recognition of processes of variation in infection that were previously obscure. This work has depended heavily on such historically important paradigms as the consideration of colony morphology in monitoring and manipulating variation. Secondly, advances in molecular biology and biotechnology have made possible the characterisation and investigation of variants at a level not previously possible, and perhaps the most important asset has been the availability of monoclonal antibodies as specific reagents for identifying, distinguishing and selecting antigenic variants.

In assembling the contributions to this volume, we have emphasised the diversity of strategies employed by micro-organisms, related to the capacity for rapid genetic variation, to enhance their potential for survival as infectious agents. The existence of alternate antigenic types among populations of pathogens (exemplified by the antigenic heterogeneity among isolates of *Escherichia coli* considered in this volume) allows for survival of the species through infection of hosts with pre-induced immune responses to prior encounters with different types of the same organism. This capacity to evade herd immunity has been recognised since serological techniques with the power to discriminate between types were developed. What has not been fully considered, and we still do not have definitive information on this, is how far diversity in organisms as the gonococcus arises from ongoing generation of variants within existing populations, perhaps by the very mechanisms we learn of in this volume. It should be emphasised here that the evaluation of variability within populations of pathogens, and attention to the clonal nature or otherwise of clinical isolates used experimentally, is of paramount importance in work in this field. A special case in considering antigenic diversity in populations is perhaps the temporal parameter in population change, best understood and exemplified by our detailed knowledge of changes in influenza virus antigenicity with time, and for this reason we have included a contribution on this topic. In addition to the occurrence of antigenic diversity within a species, the most fascinating strategy for utilization of variability is perhaps the emergence of 'new' types derived from the population of organisms present in an ongoing infectious process. The biological implications here are considerable if the process could in any way be considered as a genetic change driven by environmental change. We have no firm evidence that the latter, rather than random mutability and selection of the fittest variants, is a significant mechanism for generation of diversity, but some of the authors have begun to address this question. The remaining 'strategy' which is somewhat problematical is the ability of many organisms to lose whole constellations of virulence-related factors, and in some cases to have the potential to regain them. An example is the phase changes of *Bordetella pertussis* and, as the authors of this contribution point out, the 'biological rationale' for this capability is not understood.

The importance of variation and mechanisms of variability in the host-parasite relationship in infectious disease is now clearly established. Those systems which have now been described in detail probably represent a minority of potentially significant phenomena, and wider awareness of the resultant complexities that can be generated in pathogenesis may lead to greater understanding of many biological interactions between micro-organisms and their host species.

C.W.Penn
T.H.Birkbeck

Contributors

A.G.Barbour
National Institutes of Health, National Institute of Allergy and Infectious Diseases, Laboratory of Pathobiology, Rocky Mountain Laboratories, Hamilton, MT 59840, USA

J.E.Clements
The Johns Hopkins University, School of Medicine, Departments of Neurology and Molecular Biology and Genetics, A. Meyer Building, 600 N. Wolfe Street, Baltimore, MD 21205, USA

C.J.Duggleby
Molecular Genetics Laboratory, PHLS, CAMR, Porton Down, Salisbury, Wilts SP4 0JG, UK

A.R.Gorringe
Vaccine Research and Production Laboratory, PHLS, CAMR, Porton Down, Salisbury, Wilts SP4 0JG, UK

J.E.Heckels
Department of Microbiology, Southampton University Medical School, Southampton General Hospital, Southampton SO9 4XY, UK

C.J.Issel
Departments of Veterinary Science and Veterinary Microbiology, Louisiana State University and Louisiana Agricultural Experiment Station, Baton Rouge, LA 70803, USA

S.Kennedy-Stoskopf
The Johns Hopkins University, School of Medicine, Department of Comparative Medicine, A.Meyer Building, 600 N.Wolfe Street, Baltimore, MD 21205, USA

I.Livey
Molecular Genetics Laboratory, PHLS, CAMR, Porton Down, Salisbury, Wilts SP4 0JG, UK

R.C.Montelaro
Department of Biochemistry, Louisiana State University and Louisiana Agricultural Experiment Station, Baton Rouge, LA 70803, USA

O.Narayan
The Johns Hopkins University, School of Medicine, Departments of Neurology and Comparative Medicine, A.Meyer Building, 600 N.Wolfe Street, Baltimore, MD 21205, USA

S.Payne
Department of Biochemistry, Louisiana State University and Louisiana Agricultural Experiment Station, Baton Rouge, LA 70803, USA

R.H.A.Plasterk
Division of Biology, California Institute of Technology, Pasadena, CA 91125, USA

A.Robinson
Vaccine Research and Production Laboratory, PHLS, CAMR, Porton Down, Salisbury, Wilts SP4 0JG, UK

W.Royal,III
The Johns Hopkins University, School of Medicine, Department of Neurology, A.Meyer Building, 600 N.Wolfe Street, Baltimore, MD 21205, USA

O.Salinovich
Department of Biochemistry, Louisiana State University and Louisiana Agricultural Experiment Station, Baton Rouge, LA 70803, USA

J.R.Saunders
Department of Microbiology, University of Liverpool, Liverpool L69 3BX, UK

M.I.Simon
Division of Biology, California Institute of Technology, Pasadena, CA 91125, USA

J.J.Skehel
National Institute for Medical Research, Mill Hill, London NW7 1AA, UK

C.J.Smyth
Department of Microbiology, Moyne Institute, Trinity College, Dublin 2, Republic of Ireland

M.J.Turner
Merck Sharp & Dohme Research Laboratories, P.O.Box 2000, Rahway, NJ 07065, USA

D.C.Wiley
Department of Biochemistry and Molecular Biology, Harvard University, 7 Divinity Avenue, Cambridge, MA 02138, USA

Contents

Abbreviations

AIDS	acquired immunodeficiency syndrome
CAE	caprine arthritis–encephalitis
CL	chemiluminescence
CPE	cytopathic effects
EIAV	equine infectious anaemia virus
ELISA	enzyme-linked immunosorbent assay
FHA	filamentous haemagglutinin
HLT	heat-labile toxin
HSA	histamine-sensitizing agent
LA	leukocyte association
LPF	lymphocytosis promoting factor
LPS	lipopolysaccharide
LTR	long terminal repeat
MAb	monoclonal antibody
NA	nicotinic acid
OMP	outer membrane protein
PAGE	polyacrylamide gel electrophoresis
PBL	peripheral blood leukocytes
PMN	polymorphonuclear leukocytes
SDS	sodium dodecyl sulphate
VAT	variant antigen types
VMP	variable major proteins
VSG	variant surface glycoprotein

CHAPTER 1

Antigenic variation in the parasitic protozoa

MERVYN J.TURNER

Merck Sharp & Dohme Research Laboratories, P.O. Box 2000, Rahway, NJ 07065, USA

Introduction

The parasitic protozoa have adopted a number of survival strategies to circumvent elimination by the immune system of their host. These include adoption of an intracellular stage in the life cycle, the generation of an immunodepressed state within the host and antigenic variation. The most obvious and best studied example of antigenic variation is found in the African trypanosomes, but they are not unique amongst the parasitic protozoa in adopting this strategy. Both the *Plasmodia* and the *Babesia* have also been shown to undergo antigenic variation, although it is still by no means clear what the importance of antigenic variation is in these two species for their survival in the host. As yet, very little is known about the nature of the variant antigens expressed in these species. Much of this chapter will, therefore, concentrate on antigenic variation in the African trypanosomes, with a short section updating our knowledge on the biology and biochemistry of antigenic variation in the *Plasmodia* and *Babesia*.

Antigenic variation in the African trypanosomes

Biology

The African trypanosomes are all either cyclically transmitted by the bite of the tsetse fly (species *Glossina*) or descendents of such species which are now transmitted non-cyclically by other biting diptera or venereally. When a tsetse fly infected with *Trypanosoma brucei* takes a blood meal, metacyclic trypomastigotes are injected into the bloodstream of the mammalian host where they rapidly differentiate into the so-called long, slender trypomastigotes which replicate freely in the bloodstream by binary fission. Parasite numbers increase until an effective humoral immune response is mounted which leads to the immune destruction of more than 99% of the parasites (Seed, 1979). Some survive, however, because they have changed their antigenic profile such that they are not recognized by the immune serum. This population generates a second wave of parasitaemia which is, in turn, eliminated by the induction of a second specific immune response, but not before the precursors of a third wave of parasitaemia have been generated. This cycle can continue, apparently, indefinitely until the death of the host. At the crisis of each parasitaemia, a small fraction of the trypanosome population follows an alternative differentiation pathway leading to the generation of the so-called short, stumpy form (Mancini and Patton, 1981). This apparently does not replicate in the mammalian host (Black *et al.*, 1985), but is an essential precursor to the multiplicative stage found in the mid-gut of the tsetse fly following ingestion of the short, stumpy form in the course of a blood meal. Development in the tsetse fly is completed by a set of complex morphological and physiological changes leading, ultimately, to the appearance in the salivary glands of epimastigote forms which then develop into the metacyclic form which is again infective for the mammalian host.

All forms of the parasite found in the bloodstream of the mammalian host, and also the metacyclic trypomastigotes, are characterized by the appearance in transmission electron microscopy of an electron-dense coat covering the entire surface of the organism, some 12–15 nm thick (Vickerman, 1969). This coat is made up of a matrix of some 10^7 identical glycoprotein molecules (Cross, 1975). By the controlled expression of glycoproteins differing in their protein sequence, the trypanosome is able to change its antigenic profile and, therefore, escape immune elimination. These glycoproteins are termed variant surface glycoproteins or VSGs. Although individual trypanosomes are antigenically homogeneous, populations of trypanosomes are antigenically mixed. Thus, when infective metacyclic forms appear in the salivary glands of the tsetse fly, this population will consist of a mixture of about 20 variant antigen types (VATs) (Barry *et al.*, 1983; Crowe *et al.*, 1983). When injected into the bloodstream, these VATs are amongst the first to be detected when parasitaemia becomes patent (Barry and Emery, 1984). These metacyclic types (mVATs) constitute only a small proportion of the total VSG gene repertoire. Gene counting techniques imply the existence of more than 1000 different VSG genes (van der Ploeg *et al.*, 1982c). Hence, only about 0.1% of the repertoire is expressed in the form of mVATs.

There is some further loose ordering in the appearance of different antigenic types in the bloodstream (Gray, 1965a, 1965b; Capbern *et al.*, 1977; Kosinski, 1980). Some antigenic types, termed 'predominant', appear in the peaks of parasitaemia following clearance of the mVATs. Even when such predominant antigenic types have been cleared from the bloodstream by the host immune system, syringe passage trypanosomes isolated at a later stage of the infection into a non-immune animal frequently leads to the rapid reappearance of the predominant VATs. This is one of several observations that prove that antigenic variation is a reversible phenomenon.

Following ingestion by the tsetse fly, all antigenic types assume a common antigenicity on the loss of the surface coat in the midgut of the fly (Seed, 1964). When the parasite completes its development in the salivary glands and re-expresses the surface coat, it appears that the same set of mVATs is re-expressed as was present in the initial inoculum (Jenni, 1977a, 1977b). Thus, it is clear that the expression of metacyclic VSGs is not stochastic and must be under selective genetic or environmental control (or a combination of both). Furthermore, on re-infection of a mammalian host, the VAT ingested by the tsetse fly is always found in one of the first peaks of patent parasitaemia, suggesting some form of anamnestic response (Barry and Emery, 1984). Models for the mechanism of antigenic variation must, therefore, account for the selective activation of a subset of the total gene pool within the metacyclic population and within the predominant antigenic types plus the selective reactivation of the ingested VAT. In addition the model must explain how only one VSG at a time is expressed at the surface of the trypanosome and how expression of the surface coat is repressed and induced at different stages in the life cycle. Lastly, the model must, of course, provide a mechanism by which expression of one VSG gene can be switched to the expression of another.

Biochemistry

Characterization of VSGs has been greatly simplified by the ease with which they may be purified to homogeneity in a water-soluble form (Cross, 1975, 1984a). Using simple procedures, it is possible to prepare 50 mg amounts in 2–3 days. The purification scheme has as its basis the observation that any event leading to the rupture of the plasma membrane leads to the rapid release of the surface coat in a water-soluble form. It is this form called sVSG which has been most extensively characterized. Recently, it has become apparent that sVSG is generated by an enzymatic cleavage and, if this cleavage is inhibited, a different form of the VSG, termed the 'membrane form' or mfVSG can be isolated (Cardoso de Almeida and Turner, 1983). The nature of this cleavage reaction is described in more detail below.

Each VSG is a glycoprotein containing 450–500 amino acids and 7–17% carbohydrate by weight (reviewed in Cross, 1984b; Turner, 1984). Analysis of VSG amino acid sequences shows that the N-terminal two-thirds of the molecule is hypervariable. With the exception of the distribution of cysteines (*vide infra*), very few homologies are detectable. Within the C-terminal one-third of the molecule, however, sequence homologies do become detectable in addition to conservation of the distribution of cysteines. At the C terminus of the molecule, the homologies are so strong that it is possible to divide VSGs into subsets according to their C-terminal amino acid (Holder and Cross, 1981; Rice-Ficht *et al.*, 1981). In all cases to date, this has turned out to be aspartic acid (in one case, asparagine) or serine. Those with C-terminal aspartate are termed class I VSGs and those with serine, class II. In the purified VSG, both sVSG and mfVSG, the α-carboxyl group of the C-terminal amino acid is in an amide linkage with ethanolamine (Holder, 1983) to form part of a glycolipid which functions to anchor the VSG onto the plasma membrane of the trypanosome. In the mRNA, however, an additional 17–22 amino acids are encoded beyond the C terminus of the mature glycoprotein (Boothroyd *et al.*, 1981). These amino acids are largely hydrophobic in character and probably function to hold the precursor VSG in position until the hydrophobic tail is replaced by the glycolipid.

The hypervariable and homologous sequences can be physically separated through the action of trypsin or other proteolytic enzymes which cleave the VSG into two domains, roughly two-thirds of the way from the N terminus (Johnson and Cross, 1979). Class I VSGs usually contain eight cysteine residues in two groups of four within the homology domain and, in the case of one variant, these have been proved to form four intramolecular disulphide bonds resulting in a rather tightly folded structure (Allen and Gurnett, 1983). In one VSG, one of the groups of four cysteines has apparently been deleted with no effect on function (Cross, 1984b).

Fewer complete VSG sequences are available for class II VSGs, but they would seem to contain four or five cysteines within the C-terminal homology domain. The location of the disulphide bonds has not been established for any class II VSGs.

Within the hypervariable domain, class I VSGs generally seem to contain four, and occasionally five, cysteines at relatively conserved positions. In the one VSG in which the location of all the disulphide bonds has been established, these form two intramolecular bonds and one free sulphydryl (Allen and Gurnett, 1983). Again, insufficient data are available to make any generalizations about class II VSGs. Indeed, it should be emphasized that there are several striking exceptions to these 'rules'.

VSGs contain two types of oligosaccharide side chains (Holder and Cross, 1981). One type is N-linked to asparagine and contains N-acetylglucosamine, mannose and, occasionally, galactose. These sugars may be removed through the action of endoglycosidase H and their addition in the course of biosynthesis may be inhibited by the antibiotic, tunicamycin (McConnell *et al.*, 1982, 1983; Rovis and Dube, 1981). In all respects, these oligosaccharides are comparable with typical eukaryotic high-mannose oligosaccharides. All VSGs contain at least one such N-linked oligosaccharide side chain. In class I VSGs the site of attachment is commonly found 50 – 100 amino acids from the C terminus. Class II VSGs invariably have one such N-linked oligosaccharide attached to an asparagine four or five amino acids from the C terminus. Both class I and class II VSG may contain additional N-linked oligosaccharides. The location of such additional N-linked oligosaccharides does not seem to be particularly well conserved.

The second type of oligosaccharide side chain is altogether more interesting. As discussed briefly above, the VSG has a glycolipid covalently bound to the C-terminal amino acid and the second type of oligosaccharide forms the carbohydrate component of this glycoconjugate. Attached to the C-terminal amino acid is an ethanolamine and this is linked, possibly through a phosphodiester bond, to carbohydrate (Ferguson *et al.*, 1985b). This carbohydrate is immunogenic, in some species at least, notably rabbit, and the antibody produced cross-reacts with all VSGs. This is commonly the only cross-reaction found in comparing different VSGs and, for this reason, the carbohydrate is referred to as the 'cross-reacting determinant' or CRD (Barbet and McGuire, 1978; Cross, 1979; Holder and Cross, 1981). The principal sugar in this carbohydrate is galactose, although mannose may also be present (Holder and Cross, 1981), but the key components are a molecule of glucosamine linked to *myo*-inositol (Ferguson *et al.*, 1985b). This, in turn, is linked to phosphatidyl-*sn*-1,2-dimyristylglycerol (Ferguson and Cross, 1984; Ferguson *et al.*, 1985a).

Cleavage of the phosphodiester bond by a phosphatidylinositol-specific phospholipase C is responsible for cleavage of 1,2-dimyristyl glycerol from mfVSG to generate sVSG (Ferguson *et al.*, 1985a, 1985b). It is an intriguing question as to why a hydrophobic peptide, which could apparently function as a membrane anchor, is removed to be replaced by this unusual glycolipid, which can act as a substrate for a very interesting enzyme. A clear interpretation is that acivation of this enzyme, to act upon a substrate which is a common component of all VSGs, could provide a rapid and efficient way to release the surface coat *in vivo*. There are two stages within the life cycle at which such a mechanism might operate. One is in the midgut of the tsetse fly when the surface coat is released, apparently as sVSG, and VSG gene expression is suppressed (Overath *et al.*, 1983). The second possibility is that during the antigenic switch itself, one surface coat is released through activation of the enzyme as a second VSG is expressed.

Antigenic variation is actually rather a rare event. A single trypanosome will replicate to produce an antigenically homogeneous population in the absence of immune selection. Eventually, new antigenic variants appear, but at a level of only about one in

10^4 of the population (van Meirvenne *et al.*, 1975; Doyle, 1977). This makes it very difficult to examine an individual switching event and, hence, it has not been possible to determine the mechanics of the coat-changing operation. This appealing hypothesis for the role of the glycolipid anchor and its associated phospholipase becomes slightly less appealing in the light of recent observations that such glycolipid anchors are not confined to the VSG of trypanosomes. Such glycolipids have also been found to anchor acetylcholinesterase, 5′ nucleotidase, alkaline phosphatase, Thy-1 and p63 of *Leishmania* (reviewed in Low *et al.*, 1986). It is by no means clear, in these other examples, that there is an endogenous enzyme which acts to release the protein in water-soluble form. Obviously, we have much to learn about a novel class of membrane proteins.

The biosynthesis of VSGs has been studied in some detail. It is clear that they are synthesized as a precursor having an N-terminal hydrophobic signal peptide which is cleaved co-translationally (Boothroyd *et al.*, 1981; McConnell *et al.*, 1981). The N-linked oligosaccharides are added either co-translationally or immediately post-translationally through a mechanism that is sensitive to the presence of tunicamycin and, therefore, by implication, involves transfer of a dolichol-linked oligosaccharide (McConnell *et al.*, 1982; Rovis and Dube, 1981; Strickler and Patton, 1980). The hydrophobic tail encoded in the mRNA has never been detected directly, although its presence has been inferred from pulse–chase experiments (McConnell *et al.*, 1983). Transfer of the C-terminal glycolipid seems to take place within 1 min of VSG synthesis (Bangs *et al.*, 1985; Ferguson *et al.*, 1986). Addition of the carbohydrate component and the lipid component appears to occur concomitantly implying that the lipid is built up, perhaps on a carrier molecule, and is transferred *en bloc* to the nascent VSG.

Nevertheless, an exciting area for research in VSG biochemistry will be the identification of the biosynthetic pathway by which the glycolipid is assembled and transferred. One of the intriguing aspects of the pathway is the absolute requirement for myristic acid in the VSG. It is most unusual for acyl transferases to exhibit such preference for a particular fatty acid, and this may reflect either a very unusual enzyme or a very strict functional requirement for myristic acid in mfVSG, or both.

The extensive sequence variation seen amongst VSGs seems to provide all the explanation needed for the extraordinary antigenic diversity of these molecules. Surprisingly, however, it seems that the trypanosome only exploits a fraction of the total sequence diversity in the generation of antigenically distinct living trypanosomes. The data to support this hypothesis come from which experiments with monoclonal antibodies from which it has become clear that many monoclonal antibodies can be generated which react with the purified VSG or with acetone-fixed or air-dried films of trypanosomes in a variant-specific way, yet such monoclonal antibodies do not react with living trypanosomes (Hall and Esser, 1984; Miller *et al.*, 1984a, 1984b). These antibodies are, therefore, recognizing ‘cryptic’ antigenic determinants and such determinants would, apparently, have no role in protecting the trypanosome from the host’s immune system. Several explanations have been put forward to explain this observation. The cryptic determinants may act as decoys for the immune system by stimulating production of the irrelevant antibody, or structural constraints within the VSG itself may require sequence variation within unexposed portions of the VSG molecule as a structural counterpoise to the alterations required in exposed areas to generate new antigenic variants.

A third possibility is that trypanosomes have evolved a mechanism which allows the rapid generation of sequence diversity with VSGs, and that the first level of selection for such VSGs lies not with the ability to evade the host's immune system, but with the ability to form a compact surface coat. Once selected on this basis, the immune system would impose a second level of selection ensuring that such new variants differed antigenically in that portion of the VSG exposed at the surface of the trypanosome.

Clearly, it is important to delineate structurally between those portions of the VSG exposed at the surface of the living trypanosome and those which are buried within the surface coat. The surface coat is a very compact organelle, containing about 10^7 molecules of VSG. VSGs themselves form dimers in solution (Auffret and Turner, 1981), and at the surface of the trypanosome they may be cross-linked into much higher oligomers (Strickler and Patton, 1982), suggesting close association. Further evidence for such tight packing of the VSG comes from the observation that carbohydrate (both N-linked and lipid-linked) is not accessible on living trypanosomes, but is exposed following trypsinization (Cross and Johnson, 1976; Cardoso de Almeida, 1983). Two layers of carbohydrate have been observed by histochemical staining of thin sections in transmission electron microscopy, one layer apparently apposed to the plasma membrane, the second layer, some 5 nm above the first (Wright and Hales, 1970). This second layer is only apparent under certain conditions of fixation. Nevertheless, the simplest interpretation of these data is that this may represent the N-linked oligosaccharide whereas that apposed to the plasma membrane is the oligosaccharide component of the C-terminal glycolipid. Such tight packing of the surface coat is, presumably, necessary in order for it to fulfil its function as a barrier to the host immune system in preventing recognition of underlying common surface antigens and components of the plasma membrane, which activate complement by the alternate pathway (Tetley *et al.*, 1981). As an aside, it should be pointed out that these considerations render displacement of the surface coat, or inhibition of its correct attachment, very attractive approaches towards the chemotherapy of trypanosomiasis.

As a first step towards a complete understanding of the topography of the surface coat, a three-dimensional analysis of the structure of a VSG by X-ray crystallography has been attempted. A number of VSGs have been crystallized, at least four of which are suitable for analysis to 6 Å resolution or better (D.Wiley, personal communication). In every case, however, it is not the intact VSG which crystallizes, but a fragment. In the best studied example, this fragment comprises the N-terminal hypervariable two-thirds of the molecule, and this seems likely to be the case in other instances. To date, data for one such fragment have been collected to 6 Å resolution (Freymann *et al.*, 1984). At this level of resolution, it is not possible to see individual amino acids, rather, only the secondary structure formed by coiling such amino acids into α-helices. In this instance, about half of the amino acids appear to be in the form of α-helix. Thus, about half of two-thirds of the VSG has been visualized. Nevertheless, the arrangement of the α-helices is quite extraordinarily striking in its symmetry. The VSG appears to crystallize as a dimer in which the two monomeric subunits are related by a 2-fold axis of symmetry, parallel to the long axis of the molecule. The dimensions of the structure are such that it approximates to a cylinder of about 90 Å long by 30 Å in diameter. The long axis of the molecule contains a bundle of α-helical rods disposed around the molecular 2-fold axis of symmetry. The core of the molecule is provided by two α-helices, one in each monomer, which run virtually the full length of the molecule, turn and then run back again almost the full length of the molecule, forming two hairpin-like structures. At the 'bottom' of the molecule, four α-helices provided by the 'hairpins' interact with an additional two α-helices, each approximately 30 Å in length whose connectivity with the rest of the structure has yet to be established, but which together form a bundle with 6-fold symmetry around the molecular 2-fold axis. Beyond this region is a central core, again of approximately 30 Å in length, dominated by the four α-helices from the two hairpins and, in the upper one-third of the molecule, these helices start to diverge, executing a series of almost 180° turns, to give the molecule a 'two-headed' appearance in which each head seems to contain three α-helical segments. Of course, it is not possible at present to relate this structure to a particular orientation on the plasma membrane of the trypanosome, and we must await, at least, a higher resolution electron density map. Even so, the molecule is striking in its symmetry and it is a matter of great interest as to the extent with which this structure is conserved amongst different VSGs with widely differing primary amino acid sequences.

The variant analyzed to date is a class II VSG with a C-terminal serine. It will, therefore, be instructive both to compare this structure with the structure of a second variant of the same class, and with a class I VSG. The ultimate goal of these crystallographic studies will be to describe at higher resolution the structure of the entire VSG and to map on that structure the exposed and cryptic antigenic determinants. The longer range goal is to determine how such structures are packed in a two-dimensional array to make up the surface coat. There has been some debate as to the extent of the structure conservation within VSGs, based largely on wide differences in content of α-helices, β-pleated sheet and β-turn as either measured experimentally by circular dichroism (Duvillier *et al.*, 1983) or computed from known protein sequences using rules for secondary structure prediction (Lalor *et al.*, 1984). These measurements and calculations have been used to support the idea that VSGs will show very considerable structural diversity. It is clear, however, that calculation measures only the probability of finding a particular form of secondary structure and it cannot measure the extent to which that probability is realized. Further, the same calculations carried out on the same VSG sequence produced remarkably different results (compare Lalor *et al.*, 1984; Allen *et al.*, 1982). Thus, determination of the structure must be the last arbiter in this matter.

Molecular genetics

Antigenic variation in trypanosomes has provided a fertile field for molecular geneticists interested in the control of gene expression in simple eukaryotes (reviewed in Steinert and Pays, 1985; Boothroyd, 1985). However, as the field has progressed, it has become apparent that trypanosomes have written some of their own rules for gene expression which may not be of general applicability for higher eukaryotes. The results, of course, are no less fascinating for that. Once cDNA clones specific for VSGs became available, it rapidly became apparent that antigenic variation was associated with rearrangements amongst the VSG genes (Williams *et al.*, 1979; Hoeijmakers *et al.*, 1980). One set of observations was consistent with the appealingly simple hypothesis that, in order to be expressed, a VSG gene was duplicated from a storage copy of the gene (the basic copy or BC), and the duplicated gene was transposed to a unique expression site associated with a telomere (Hoeijmakers *et al.*, 1980; Payes *et al.*, 1981a, 1981b; De Lange and Borst, 1982; Williams *et al.*, 1982; Bernards *et al.*, 1981). Displacement of the gene previosuly occupying this expression site by the incoming gene would provide the mechanism by which unique transcription of a single VSG gene was maintained. However, many VSG genes are already resident at telomeres and can be activated with or without a duplicative transposition (Pays *et al.*, 1983a; Laurent *et al.*, 1984; Bernards *et al.*, 1984; Majiwa *et al.*, 1982). Furthermore, both restriction enzyme mapping and pulse field gradient electrophoresis imply the existence of more than one expression site with a minimum of three (Longacre *et al.*, 1983; Myler *et al.*, 1984a; van der Ploeg *et al.*, 1984a, 1984b). Identification of a promoter associated with VSG gene expression has proved extremely elusive, partly because transcription seems to be initiated a long way upstream of the VSG gene (Pays *et al.*, 1982) and partly because the trypanosome employs discontinuous synthesis involving either a bimolecular splicing mechanism or RNA-primed transcription to generate the VSG mRNA (Campbell *et al.*, 1984a, 1984b; Kooter *et al.*, 1984; Milhausen *et al.*, 1984).

The main features of VSG gene transcription may be summarized as follows. The trypanosome genome contains perhaps 1000 VSG genes (van der Ploeg *et al.*, 1982c). These may be found in two different environments—chromosome-internal and telomeric. In order to be expressed, chromosome-internal genes must be duplicated and the duplicate transposed to a telomere. This duplicative transposition is brought about by a gene conversion (Pays *et al.*,1983b) in which the converted DNA segment is usually about 3 kb in length, about half of which comprises the VSG gene and the remainder constituting a co-transposed sequence, most of which lies to the 5′ side of the gene (Michaels *et al.*, 1983; Laurent *et al.*, 1983; Pays *et al.*, 1983a, 1983b; van der Ploeg *et al.*, 1982a). This donor 3-kb segment is bracketed by repetitive sequences which are different on the 5′ and the 3′ sides, and which are believed to be involved in the recombination to ensure correct orientation of the gene when transposed to the telomeric expression site (Pays *et al.*, 1982; Campbell *et al.*, 1984b; Lieu *et al.*, 1983). Some basic copy genes are already telomeric in nature, but may be duplicated by the same mechanism of gene conversion to give an additional telomeric copy. Although gene conversion frequently involves transposition of a 3-kb element, gene conversions involving replacment of very much longer sequences (up to 40 kb) have been observed (Pays *et al.*, 1983a). Furthermore, examples of partial gene conversions have been found in which the sequence replacement terminates within the VSG gene itself (Pays *et al.*, 1983b). Under such circumstances, a hybrid VSG gene is formed, containing sequence elements of the donor and recipient in the gene conversion and this may provide a mechanism by which sequence diversity can be generated within the VSG gene pool.

Once at a telomere, a VSG gene may be transcribed. Transcriptionally active VSG genes may be distinguished by their sensitivity to DNase I, indicative of the characteristic open conformation of chromatin (Pays *et al.*, 1981a). Sequence analysis of the VSG mRNA compared with the sequence of the gene established that the VSG mRNA has, at its 5′ terminus, a sequence of 34 bases which are absent from the corresponding location in the gene (Boothroyd and Cross, 1982). This sequence, known as the mini-exon, implied that a splicing event was required for the generation of a mature mRNA (van der Ploeg *et al.*, 1982b). Early attempts to map the 35-base sequence in relation to the expressed VSG gene were unsuccessful insofar as the two sequences appeared to be separated by at least 50 kb (De Lange *et al.*, 1983; Nelson *et al.*, 1983). Such data suggested a very large precursor mRNA and, indeed, large VSG mRNA transcripts have been detected. However, it soon became clear that the VSG mRNA synthesis is discontinuous, as a 137-base transcript containing the 34-base sequence may readily be detected (Campbell *et al.*, 1984a; Kooter *et al.*, 1984; Milhausen *et al.*, 1984). Also, pulse field gradient electrophoresis established that the 34-base sequence was, in some instances, located on a chromosome other than that from which the VSG was being transcribed (van der Ploeg *et al.*, 1984a, 1984b; Guyaux *et al.*, 1985). It would seem that either a bimolecular splicing event is needed to generate full length VSG mRNA, or that the transcript containing the 34-base sequence acts to prime transcription from a site upstream of the VSG gene. It has not yet proved possible to identify the transcriptional start point for the VSG gene. Interestingly, the polymerase responsible for transcription of the 34-base sequence appears to be relatively insensitive to α-amanitin and transcription of the VSG gene itself is extremely insensitive, in comparison with the effect on transcription of housekeeping genes or tubulin (Kooter and Borst, 1984). VSG gene transcription may, therefore, involve an entirely novel class of RNA polymerase.

The 34-base sequence is, itself, found within a larger sequence of 1.1 kb which is tandemly repeated some 200 times within the genome (De Lange *et al.*, 1983; Nelson *et al.*, 1983). Although the original hypothesis that this block of sequences acted as a strong promoter for VSG gene transcription was an attractive one, it transpired that many if not all mRNAs in trypanosomes (De Lange *et al.*, 1984b; Parsons *et al.*, 1984), and indeed in many other kinetoplastida which do not undergo antigenic variation (De Lange *et al.*, 1984a; Nelson *et al.*, 1984), are capped by this sequence of 34 bases. Hence, discontinuous transcription may be the rule in kinetoplastida and not the exception in trypanosomes.

Although it is essential for a VSG gene to be placed at a telomere before it may become transcriptionally active, once at a telomere, it can be inactivated in at least two, and possibly three, different ways (Pays *et al.*, 1985). Firstly, it may act as the recipient in a gene conversion event and be displaced from the genome. Secondly, it may be involved in a reciprocal recombination event involving telomere exchange between active and inactive genes. Thirdly, it may be inactivated with no apparent gene rearrangement. Similarly, a telomeric-inactive VSG gene may be activated with no evidence for either telomere translocation or gene conversion. Attempts to define differences between such active and inactive telomeres have been unsuccessful. Both 5′ and 3′ to the VSG gene are so-called 'barren regions'—sequences which vary in length from 5 to 15 kb—relatively devoid of restriction enzyme sites. Sequence analyses of such 5′ barren regions have shown that they are comprised of a relatively AT-rich repeating element which is identical in both expressed and inactive VSG genes (Campbell *et al.*, 1984b; Young, 1985). The 3′ barren region sequence is less well characterized, but in *Trypanosoma equiperdum* it seems to contain a modified cytosine residue, the function of which is unclear (Raibaud *et al.*, 1983).

Trypanosomes have many chromosomes, perhaps as many as 100, some of which are very small (<100 kb) (Williams *et al.*, 1982; van der Ploeg and Cornelissen, 1984). Thus, many VSG genes have the opportunity to occupy a telomeric site. It remains a conundrum how transcription of telomeric VSG genes is controlled so that only one VSG at a time is expressed at the surface of the trypanosome. The telomeric environment may favour early expression of a VSG gene, and this may provide at least part of the explanation for the phenomenon of predominance (Laurent *et al.*, 1983; Myler *et al.*, 1984b), and also the selective expression of mVATs. In addition, it is known that when a trypanosome ceases expression of a VSG gene on entry into the midgut of the tsetse fly or in the tissue culture equivalent, transcription stops within 30 min and the telomere becomes inactive (Overath *et al.*, 1983). However, the VSG gene is retained at the telomere rather than being subjected to gene conversion into oblivion and so may be preferentially expressed following the completion of the life cycle in the tsetse fly, and this may provide the explanation for the anamnestic response.

In summary, it may be said that although the study of the molecular genetics of antigenic variation in trypanosomes has been an exciting and fast moving field, and tremendous progress has been made, some major questions remain unanswered.

Antigenic variation in malaria

Although antigenic variation in malaria is well established experimentally, its relevance to the parasite for its survival in the immunized host has been a matter of some contention. Antigenic diversity within different species of *Plasmodia* and within antigens expressed at the same stage of the life cycle is generally thought to be a consequence of genotypic diversity, that is the circulation of different alleles amongst the population (reviewed in Hommel, 1985). However, phenotypic antigenic variation, i.e. the expression of different antigen gene products by an individual parasite, has been documented for one antigen only, the so-called SICA antigen, named after the assay (schizont-infected cell agglutination) in which it was discovered (Brown and Brown, 1965).

Schizonts are produced in the erythrocytic stage of the life cycle of *Plasmodia*. When an infected mosquito takes a blood meal, sporozoites injected into the bloodstream rapidly find their way to the parenchymal cells of the liver where they undergo a cycle of schizogony. Released merozoites invade red cells and enter a second cycle of schizogony which is perpetuated by continued relase of merozoites and subsequent re-invasion of red cells. This constitutes the erythrocytic stage. In each cycle, a small proportion of gametocytes is formed. Sexual reproduction occurs within the mosquito gut to form oocytes which ultimately produce sporozoites to complete the life cycle.

The first convincing evidence for the existence of antigenic variation during the erythrocytic cycle came in a series of experiments in which control by drug treatment of acute infections of rhesus monkeys with *P. knowlesi* produced a chronic infection characterized by the sequential appearance of antigenically distinct populations. These populations were identified by agglutination of schizont-infected red cells (but not of uninfected red cells or immature schizonts) by infection serum in a variant-specific fashion (Brown and Brown, 1965; Brown *et al.*, 1968).

Although this work was done on uncloned populations, which apparently left only the possibility of genotypic variation, incubation of as few as 10 schizont-infected red blood cells with agglutinating antibody led to the appearance of a new and antigenically distinct variant, strongly suggestive of phenotypic rather than genotypic change (Brown,

1973). Subsequently, this work has been reproduced using two cloned populations of *P. knowlesi* in which one clone was produced from the other by immune selection in a monkey (Howard *et al.*, 1983; Howard, 1984). The variant antigen was identified by surface labelling, followed by immunoprecipitation after detergent extraction of schizont-selected red blood cells. Many common components were detected between the two clones, but each expressed a variant-specific antigen having slightly different molecular weights in the 200 000 range. The antigens seemed to be of parasite origin, since they could be labelled with [^{35}S]methionine during parasite growth.

Antigenic variation has also been observed in schizont-infected red cells during infection with *P. falciparum* (Hommel *et al.*, 1983). As yet, the biochemical nature of the antigens has not been established. The spleen seems to have an important role in determining the expression of the SICA antigen, both in *P. knowlesi* and *P. falciparum*. Thus, in *P. falciparum*, passage of a cloned population of schizont-infected red cells, through a splenectomized host, led to phenotypic change in the variant antigen expressed (Hommel *et al.*, 1983). Transfer of the parasite from a splenectomized to an intact animal, led to a reappearance of the original phenotype. The role of the spleen in this switch is not understood.

As yet, no defined function has been produced for the variant antigens of *Plasmodia*. Unlike VSGs of trypanosomes, the variant antigen seems to be present in very small amounts, and is clearly not the principal mechanism of immune evasion. Indeed, it may be asked, why does an intra-erythrocytic parasite which occupies a parasitopherous vacuole need to express a parasite-derived product on the surface of the host cell, a process which requires export across two sets of plasma membrane.

The variant antigens appear to be strain-specific, that is different strains express different repertoires of variant antigens (discussed in Howard, 1984; Howard *et al.*, 1983). One strain-specific antigen of *P. falciparum* has been implicated in attachment of trophozoite and schizont-infected red blood cells to the capillary and venular endothelium (Udeinya *et al.*, 1983; Leech *et al.*, 1984). This antigen may prevent infected erythrocytes from circulating through the spleen where they could be eliminated. Such an antigen could function like the influenza virus haemagglutinin molecule which retains its function of binding to the target cell whilst undergoing its extensive sequence variation to generate antigenically distinct variants (J.Skehel, Chapter 2). At present, however, there is no evidence to link the variant antigen of schizont-infected red blood cells with the endothelial binding antigen.

Antigenic variation in Babesia

Members of the genus *Babesia* produce a tick-borne disease resembling malaria in all domestic animals. The disease is of major economic importance in ruminants, producing symptoms of fever, anaemia, haemoglobinuria and jaundice. When infected ticks feed, parasites are injected into the bloodstream and immediately enter an asexual cycle of erythrocytic schizogony. Recovery correlates with disappearance of parasite from the bloodstream, but in the case of *B. bovis* at least, subpatent infections can continue for up to 4 years, and in both *B. bovis* and *B. rodhaini* the development of chronic infection appears to correlate with the appearance of antigenically distinct populations which can be detected by variant-specific agglutination of infected red cells (Curnow,

1973; Phillips, 1971). Antigenic variation has been generated from cloned populations (Roberts and Tracey-Patte,1975), but inter-conversion of antigenically-distinct cloned variants which would establish true phenotypic variation has not been demonstrated in a laboratory model. However, in a large-scale experiment in cattle, there did seem to be apparent reversion to the basic antigenic type following cyclical transmission of *B. bovis* (Curnow,1983). Unlike the variant antigen of the *Plasmodia*, there seems to be no evidence for an effect on this antigen in *Babesia* by passage in splenectomized animals (Curnow, 1973), a treatment which has produced an attenuated strain of *B. bovis* which is used for the production of a live vaccine (Callow and Mellors, 1966).

As part of an approach to identify babesial antigens that induce protective immunity, and which might form the basis of a subunit vaccine, a study has been performed on the protein antigens of an attenuated non-virulent strain of *B. argentina,* on a virulent line from the same strain and a second virulent line from a different strain (Kahl *et al.*, 1982). Proteins in parasitized erythrocytes were metabolicaly labelled and were characterized by detergent extraction and gel electrophoresis in the presence or absence of immune serum. Although marked similarities were observed in the repertoire of proteins labelled, differences were observed in highly acidic, dominantly labelled polypeptides of molecular weight approximately 40 000. No amino acid homologies between these proteins could be established by peptide mapping, and these proteins were, therefore, considered as candidates for the variant antigen. However, these experiments were not performed with cloned populations and it is not known whether the antigens are strain-specific, produced by genotypic variation or whether they do, indeed, represent the antigens of different phenotypes.

Conclusion

Antigenic variation represents a very potent survival strategy for parasites. It appears to have been embraced wholeheartedly by the African trypanosomes, whereas the other parasitic protozoa in which this mechanism has been found to operate seem to prefer to use it as just one weapon in a well-stocked arsenal. It is clear that the diversity of antigenic variation in the African trypanosomes is such that the chances of producing a subunit vaccine are remote indeed. Within the other parasitic protozoa, it is by no means clear that this is the case. Indeed, much excitement has been generated over the last few years at the prospect of producing a subunit vaccine for malaria. Whether phenotypic antigenic variation is likely to prove an obstacle to this development has yet to be seen. Indeed it may be that genotypic variation will post a more substantial difficulty. Many malarial antigens being evaluated as candidates for a subunit vaccine contain immunodominant repeated polypeptide sequences (reviewed in Kemp *et al.*, 1986). Whilst such molecules may prove potent immunogens, the speed with which a point mutation within one repeat could spread through all the repeats by unequal crossing-over to generate an antigenically distinct variant, should not be underestimated. The parasitic protozoa have been around at least as long as we have, and may have sufficient resources to outwit even the most cunning gene cloner!

References

Allen,G. and Gurnett,L.P. (1983) Location of the six disulphide bonds in a variant glycoprotein (VSG 117) from *Trypanosoma brucei. Biochem. J.*, **209**, 481–487.

Allen,G., Gurnett,L.P. and Cross,G.A.M. (1982) Complete amino acid sequence of a variant surface glycoprotein (VSG 117) from *Trypanosoma brucei. J. Mol. Biol.*, **157**, 527–546.

Auffret,C.A. and Turner,M.J. (1981) Variant specific antigens of *Trypanosoma brucei* exist in solution as glycoprotein dimers. *Biochem. J.*, **143**, 647–650.

Bangs,J.D., Hereld,D., Krakow,J.L., Hart,G.W. and Enghard,P.T. (1985) Rapid processing of the carboxyl terminus of a trypanosome variant surface glycoprotein. *Proc. Natl. Acad. Sci. USA*, **82**, 3207–3211.

Barbet,A.F. and McGuire,T.C. (1978) Cross-reacting determinants in variant-specific surface antigens of African trypanosomes. *Proc. Natl. Acad. Sci. USA*, **75**, 1989–1993.

Barry,J.D. and Emery,D.L. (1984) Parasite development and host responses during the establishment of *Trypanosoma brucei* infection transmitted by tsetse fly. *Parasitology*, **88**, 67–84.

Barry,J.D., Crowe,J.S. and Vickerman,K. (1983) Instability of the *Trypanosoma brucei rhodesiense* metacyclic variable antigen repertoire. *Nature*, **306**, 699–701.

Bernards,A., van der Ploeg,L.H.T., Frasch,A.C.C., Borst,P., Boothroyd,J.C. and Cross,G.A.M. (1981) Activation of trypanosome surface glycoprotein genes involves the duplication-transposition leading to an altered 3′ end. *Cell*, **27**, 497–505.

Bernards,A., De Lange,T., Michaels,P.A.M., Lieu,A.Y.C., Huisman,M.J. and Borst,P. (1984) Two modes of activation of a single surface antigen gene of *Trypanosoma brucei*. *Cell*, **36**, 163–170.

Black,S.J., Sendashonga,C.N., O'Brien,C., Borowy,N.K., Naessens,M., Webster,P. and Murray,M. (1985) Regulation of parasitemia in mice infected with *Trypanosoma brucei*. *Curr. Top. Microbiol. Immunol.*, **117**, 93–118.

Boothroyd,J.C. (1985) Antigenic variation in African trypanosomes. *Annu. Rev. Microbiol.*, **39**, 475–502.

Boothroyd,J.C. and Cross,G.A.M. (1982) Transcripts coding for variant surface glycoproteins of *Trypanosoma brucei* have a short identical exon at their 5′ end. *Gene*, **20**, 281–289.

Boothroyd,J.C., Paynter,C.A., Cross,G.A.M., Bernards,A. and Borst,P. (1981) Variant surface glycoproteins of *Trypanosoma brucei* are synthesized with cleavable, hydrophobic sequences at the carboxy and amino termini. *Nucleic Acids Res.*, **9**, 4743–4745.

Brown,I.N., Brown,K.N. and Hills,L.A. (1968) Immunity to malaria: The antibody response to antigenic variation by *Plasmodium knowlesi*. *Immunology*, **14**, 127–138.

Brown,K.N. (1973) Antibody induced variation in malaria parasites. *Nature*, **242**, 49–50.

Brown,K.N. and Brown,I.N. (1965) Immunity to malaria: Antigen variation in chronic infections of *P. knowlesi*. *Nature*, **208**, 1286–1288.

Callow,L.L. and Mellors,L.T. (1966) A new vaccine for *B. argentina* infection prepared in splenectomized calves. *Aust. Vet. J.*, **42**, 464–465.

Campbell,D.A., Thornton,D.A. and Boothroyd,J.C. (1984a) Apparent discontinuous transcription of *Trypanosoma brucei* in variant surface antigen genes. *Nature*, **311**, 350–355.

Campbell,D.A., Van Bree,M.P. and Boothroyd,J.C. (1984b) The 5′ limit of transposition and upstream barren region of a trypanosome VSG gene: tandem 76 base pair repeats flanking $(TAA)_{90}$. *Nucleic Acids Res.*, **12**, 2759–2774.

Capbern,A., Giroud,C., Baltz,T. and Mattern,P. (1977) *Trypanosoma equiperdum:* etude des variations antigenique au cours de la trypanosomose experimentale du lapin. *Exp. Parasitol.*, **42**, 6–13.

Cardoso de Almeida,M.L. (1983) Mode of attachment of VSGs to the plasma membrane of *T. brucei*. Ph.D. Thesis, University of Cambridge.

Cardoso de Almeida,M.L. and Turner,M.J. (1983) The membrane form of variant surface glycoproteins of *Trypanosoma brucei*. *Nature*, **302**, 349–352.

Cross,G.A.M. (1975) Identification, purification and properties of clone-specific glycoprotein antigens constituting the surface coat of *Trypanosoma brucei*. *Parasitology*, **71**, 393–417.

Cross,G.A.M. (1979) Cross-reacting determinants in the C-terminal regions of trypanosome variant surface antigens. *Nature*, **277**, 310–312.

Cross,G.A.M. (1984a) Release and purification of *Trypanosoma brucei* variant surface glycoprotein. *J. Cell. Biochem.*, **24**, 59–90.

Cross,G.A.M. (1984b) Structure of the variant glycoproteins and surface coat of *Trypanosoma brucei*. *Phil. Trans. R. Soc. Lond., Ser. B*, **307**, 3–12.

Cross,G.A.M. and Johnson,J.G. (1976) Structure and organization of the variant specific surface antigens of *Trypanosoma brucei*. In *Biochemistry of Parasites and Host-Parasite Relationships. Proceedings of the Second International Symposium on the Biochemistry of Parasite and Host-Parasite Relationships.* Van den Bossche,H. (ed.), North Holland Publishing, Amsterdam, pp. 413–420.

Crowe,J.S., Barry,J.D., Luckins,A.G., Ross,C.A. and Vickerman,K. (1983) All metacyclic variable antigen types of *Trypanosoma congolense* identified using monoclonal antibodies. *Nature,* **306**, 389–391.

Curnow,J.A. (1973) Studies on antigenic changes and strain differences in *Babesia argentina* infections. *Aust. Vet. J.,* **49**, 279–283.

De Lange,T. and Borst,P. (1982) Genomic environment of the expression-linked extra copies of genes for surface antigens of *Trypanosoma brucei* resembles the end of a chromosome. *Nature,* **299**, 451–453.

De Lange,T., Lieu,A.Y.C., van der Ploeg,L.H.T., Borst,P., Trump,M.C. and Van Boom,J.A.H. (1983) Tandem repetition of the 5′ mini exon of variant surface glycoprotein genes: a multiple promoter for VSG gene transcription? *Cell,* **34**, 891–900.

De Lange,T., Berkvens,T.N., Veeneman,H.J., Frasch,A.C., Barry,J.D. and Borst,P. (1984a) The common 5′ terminal sequence of messenger RNAs in three trypanosome species. *Nucleic Acids Res.,* **12**, 4431–4443.

De Lange,T., Michaels,P.A., Veeneman,H.J., Cornelissen,A.W. and Borst,P. (1984b) Many trypanosome messenger RNAs share a common 5′ terminal sequence. *Nucleic Acids Res.,* **12**, 3777–3790.

Doyle,J.J. (1977) Antigenic variation in the salivarian trypanosomes. In *Immunity to Blood Parasites of Animals and Man.* Miller,L.H., Pino,J.A. and McKelvey,J.J. (eds), Plenum Press, London and New York, pp. 31–63.

Duvillier,G., Aubert,J.P., Baltz,T., Richet,C. and Degand,P. (1983) Variant specific surface antigens from *Trypanosoma equiperdum:* chemical and physical studies. *Biochem. Biophys. Res. Commun.,* **110**, 491–498.

Ferguson,M.A.J. and Cross,G.A.M. (1984) Myristylation of the membrane form of a *Trypanosoma brucei* variant surface glycoprotein. *J. Biol. Chem.,* **259**, 3011–3015.

Ferguson,M.A.J., Haldar,K. and Cross,G.A.M. (1985a) *Trypanosoma brucei* variant surface glycoprotein has a SN-1,2-dimyristyl glycerol membrane anchor at its COOH terminus. *J. Biol. Chem.,* **260**, 4963–4968.

Ferguson,M.A.J., Low,M.G. and Cross,G.A.M. (1985b) Glycerol-*sn*-1,2-dimyristyl phosphatidylinositol is covalently linked to *Trypanosoma brucei* variant surface glycoprotein. *J. Biol. Chem.,* **260**, 14547–14555.

Ferguson,M.A.J., Duszenko,M., Lamont,G.S., Overath,P. and Cross,G.A.M. (1986) Biosynthesis of *Trypanosoma brucei* VSG. N-glycosylation and addition of a phosphatidylinositol membrane anchor. *J. Biol. Chem.,* in press.

Freymann,D., Metcalf,P., Turner,M.J. and Wiley,D. (1984) The six Å resolution X-ray structure of a variable surface glycoprotein from *Trypanosoma brucei*. *Nature,* **311**, 167–169.

Gray,A.R. (1965a) Antigenic variation in clones of *Trypanosoma brucei*. I. Immunological relationships of the clones. *Ann. Trop. Med. Parasitol.,* **59**, 27–36.

Gray,A.R. (1965b) Antigenic variation in a strain of *Trypanosoma brucei* transmitted by *Glossina morsitans* and *G.palpalis. J. Gen. Microbiol.,* **41**, 195–214.

Guyaux,M., Cornelissen,A.W., Pays,E., Steinert,M. and Borst,P. (1985) *Trypanosoma brucei*: a surface antigen mRNA discontinuously transcribed from two distinct chromosomes. *EMBO J.,* **4**, 995–998.

Hall,T. and Esser,K. (1984) Topologic mapping of protective and non-protective epitopes on the variant surface glycoprotein of the wRAT at 1 clone of *Trypanosoma brucei rodesiense. J. Immunol.,* **132**, 2059–2063.

Hoeijmakers,J.H.J., Frasch,A.C.C., Bernards,A., Borst,P. and Cross,G.A.M. (1980) Novel expression-linked copies of the genes for variant surface antigens in trypanosomes. *Nature,* **284**, 78–80.

Holder,A.A. (1983) Carbohydrate is linked through ethanolamine to the C-terminal amino acid of *Trypanosoma brucei* variant surface glycoprotein. *Biochem. J.,* **209**, 261–262.

Holder,A.A. and Cross,G.A.M. (1981) Glycopeptides from variant surface glycoprotein of *Trypanosoma brucei*. C-terminal location of antigenically cross-reacting carbohydrate moieties. *Mol. Biochem. Parasitol.,* **2**, 135–150.

Hommel,M. (1985) Antigenic variation in malaria parasites. *Immunol. Today,* **6**, 28–33.

Hommel,M., David,P.H. and Oligino,L.B. (1983) Surface alterations of erythrocytes in *Plasmodium falciparum* malaria. *J. Exp. Med.,* **157**, 1137–1148.

Howard,R.J. (1984) Antigenic variation of blood stage malaria parasites. *Phil. Trans. R. Soc. Lond., Ser. B,* **307**, 141–158.

Howard,R.J., Barnwell,J.W. and Kao,V. (1983) Antigenic variation in *Plasmodium knowlesi* malaria: Identification of the variant antigen on infected erythrocytes. *Proc. Natl. Acad. Sci. USA,* **80**, 4129–4133.

Jenni,L. (1977a) Comparison of antigenic types of *Trypanosoma* (T.) *brucei* strins transmitted by *Glossina* M. *morsitans*. *Acta Trop.*, **34**, 35–41.

Jenni,L. (1977b) Antigenic variation in cyclically transmitted strains of the *T. brucei* complex. *Ann. Soc. Belge Med. Trop.*, **57**, 383–386.

Johnson,J.G. and Cross,G.A.M. (1979) Selective cleavage of variant surface glycoproteins from *Trypanosoma brucei*. *Biochem. J.*, **178**, 689–697.

Kahl,L.P., Anders,R.F., Rodwell,B.J., Timms,P. and Mitchell,G.F. (1982) Variable and common antigens of *Babesia bovis* parasites differing in strain and virulence. *J. Immunol.*, **129**, 1700–1705.

Kemp,D.J., Coppel,R.L., Stahl,H.D., Bianco,A.E., Corcoran,L.M., MacIntyre,P., Langford,C.J., Favaloro,J.N., Crewther,P.E., Brown,G.V., Mitchell,G.F. and Anders,R.F. (1986) Genes for antigens of *Plasmodium falciparum*. *Parasitology,* in press.

Kooter,J.M. and Borst,P. (1984) Alpha amanitin insensitive transcription of variant surface glycoprotein genes provides further evidence for discontinuous transcription in trypanosomes. *Nucleic Acids Res.*, **12**, 9457–9472.

Kooter,J.M., De Lange,T. and Borst,P. (1984) Discontinuous synthesis of mRNA in trypanosomes. *EMBO J.*, **3**, 2387–2392.

Kosinski,R.J. (1980) Antigenic variation in trypanosomes: a computer analysis of variant order. *Parasitology,* **80**, 343–357.

Lalor,T.M., Kjeldgaard,M., Shimamoto,G.T., Strickler,J.E., Konigsberg,W.H. and Richards,F.F. (1984) Trypanosome variant-specific glycoproteins: a polygene protein family with multiple folding patterns. *Proc. Natl. Acad. Sci. USA,* **81**, 998–1002.

Laurent,M., Pays,E., Delinte,K., Magnus,E., Van Meirvenne,N. and Steinert,M. (1984) Evolution of a trypanosome surface antigen gene repertoire linked to non-duplicative gene activation. *Nature,* **308**, 370–373.

Laurent,N., Pays,E., Magnus,E., Van Meirvenne,N., Matthyssens,G., Williams,R.O. and Steinert,M. (1983) DNA rearrangements linked to the expression of a predominant surface antigen gene of trypanosomes. *Nature,* **302**, 263–266.

Leech,J.H., Barnwell,J.W., Miller,L.H. and Howard,R.J. (1984) Identification of a strain-specific malarial antigen exposed on the surface of *Plasmodium falciparum* infected erythrocytes. *J. Exp. Med.,* **159**, 1567–1575.

Liu,A.Y.C., van der Pleog,L.H.T., Rijsewijk,F.A.M. and Bort,P. (1983) The transposition unit of variant surface glycoprotein gene 118 of *Trypanosoma brucei*: presence of repeated elements at its border and absence of promoter associated sequences. *Nucleic Acids Res.*, **10**, 593–609.

Longacre,S., Hibner,U., Raibaud,A., Eisen,H., Baltz,T., Giroud,C. and Baltz,B. (1983) DNA rearrangements and antigenic variation in *Trypanosoma equiperdum:* multiple expression-linked sites in independent isolates of trypanosomes expressing the same antigen. *Mol. Cell. Biol.*, **3**, 399–409.

Low,M.G., Ferguson,M.A.J., Futerman,A.H. and Silman,I. (1986) Covalently attached phosphatidylinositol as a hydrophobic anchor for membrane proteins. *Trends Biochem. Sci,* in press.

Majiwa,P.A.O., Young,J.R., Englund,P.T., Shapiro,S. and Williams,R.O. (1982) Two distinct forms of surface antigen gene rearrangement in *Trypanosoma brucei*. *Nature,* **297**, 514–516.

Mancini,P.E. and Patton,C.L. (1981) Cyclic 3′, 5′ adenosine monophosphate levels during the developmental cycle of *Trypanosoma brucei brucei* in the rat. *Mol. Biochem. Parasitol.*, **3**, 19–31.

McConnell,J., Gurnett,A.M., Cordingley,J.S., Walker,J.E. and Turner,M.J. (1981) Biosynthesis of *Trypanosoma brucei* variant surface glycoprotein. I. Synthesis, size and processing of an N-terminal signal peptide. *Mol. Biochem. Parasitol.*, **4**, 225–242.

McConnell,J., Cordingley,J.S. and Turner,M.J. (1982) Biosynthesis of *Trypanosoma brucei* variant surface glycoproteins — *in vitro* processing of signal peptide and glycosylation using heterologous rough endoplasmic reticulum vesicles. *Mol. Biochem. Parasitol.*, **6**, 161–174.

McConnell,J., Turner,M.J. and Rovis,L. (1983) Biosynthesis of *Trypanosoma brucei* variant surface glycoproteins — analysis of carbohydrate heterogeneity and timing of post-translational modifications. *Mol. Biochem. Parasitol.*, **8**, 119–135.

Miochaels,P.A.M., Lieu,A.Y.C., Bernards,A., Sloof,P., Van der Bijl,N.M.W., Schinkel,A.H., Menke,H.H., Borst,P., Veeneman,G.H., Trump,N.C. and Van Boom,J.H. (1983) Activation of the genes for variant surface glycoproteins 117 and 118 in *Trypanosoma brucei*. *J. Mol. Biol.*, **166**, 537–556.

Milhausen,M., Nelson,R.G., Sather,S., Selkirk,M. and Agabian,N. (1984) Identification of a small RNA containing the trypanosome spliced leader: a donor of shared 5′ sequences of *Trypanosomatidae* mRNAs? *Cell,* **38**, 721–729.

Miller,E.N., Allan,L.M. and Turner,M.J. (1984a) Topological analysis of antigenic determinants of a variant surface glycoprotein of *Trypanosoma brucei*. *Mol. Biochem., Parasitol.*, **13**, 67–81.
Miller,E.N., Allan,L.M. and Turner,M.J. (1984b) Relationship of antigenic determinants to structure within a variant surface glycoprotein of *Trypanosoma brucei*. *Mol. Biochem., Parasitol.*, **13**, 309–322.
Myler,P., Nelson,R.G., Agabian,N. and Stuart,K. (1984a) Two mechanisms of expression of a predominant variant antigen gene of *Trypanosoma brucei*. *Nature*, **309**, 282–284.
Myler,P.J., Allison,J., Agabian,N. and Stewart,K. (1984b) Antigenic variation in African trypanosomes by gene replacement or activation of alternate telomeres. *Cell*, **39**, 203–211.
Nelson,R.G., Parsons,M., Barr,P.J., Stuart,K., Selkirk,M. and Agabian,N. (1983) Sequences homologous to variant antigen mRNAs spliced leader are located in tandem repeats and variable orphons in *Trypanosoma brucei*. *Cell*, **34**, 901–909.
Nelson,R.G., Parson,M., Selkirk,M., Newport,G., Barr,P.J. and Agabian,N. (1984) Sequences homologous to variant antigen mRNA spliced leader in *Trypanosomatidae* which do not undergo antigenic variation. *Nature*, **308**, 665–667.
Overath,P., Czichos,J., Stock,V. and Nonnengaesser,C. (1983) Repression of glycoprotein synthesis and release of surface coat during transformation of *Trypanosoma brucei*. *EMBO J.*, **2**, 1721–1728.
Parsons,M., Nelson,R.G., Watkins,K.P. and Agabian,N. (1984) Trypanosome mRNAs share a common 5′ spliced leader sequence. *Cell*, **38**, 309–316.
Pays,B., Guyaux,N., Aerts,D., Van Meirvenne,N. and Steinert,M. (1985) Telomeric reciprocal recombination as a possible mechanism for antigenic variation in trypanosomes. *Nature*, **316**, 562–564.
Pays,E., Lheureux,M. and Steinert,M. (1981a) The expression-linked copy of the surface antigen gene in *Trypanosoma* is probably the one transcribed. *Nature*, **292**, 265–267.
Pays,E., Van Meirvenne,N., La Ray,D. and Steinert,M. (1981b) Gene duplication and transposition linked to antigenic variation in *Trypanosoma brucei*. *Proc. Natl. Acad. Sci. USA*, **78**, 2673–2677.
Pays,E., Lheureux,M. and Steinert,M. (1982) Structure and expression of a *Trypanosoma brucei gambiense* variant specific antigen gene. *Nucleic Acids Res.*, **10**, 3149–3163.
Pays,E., Delauw,M.-S., Van Assel,S., Laurent,M., Vervoort,T., Van Meirvenne,N. and Steinert,M. (1983a) Modifications of a *Trypanosoma b. brucei* antigen gene repertoire by different DNA recombinational mechanisms. *Cell*, **35**, 721–731.
Pays,E., Van Assel,S., Laurent,M., Darville,M., Vervoort,T., Van Meirvenne,N. and Steinert,M. (1983b) Gene conversion as a mechanism for antigenic variation in trypanosomes. *Cell*, **34**, 371–381.
Phillips,R.S. (1971) Antigenic variation in *Babesia rodhaini* demonstrated by immunization with irradiated parasites. *Parasitology*, **63**, 315–322.
Raibaud,A., Gaillard,C., Longacre,S., Hibner,U., Buck,G., Bernardi,G. and Eisen,H. (1983) Genomic environment of variant surface antigen genes of *Trypanosoma equiperdum*. *Proc. Natl. Acad. Sci. USA*, **80**, 4306–4310.
Rice-Ficht,A.C., Chen,K.K. and Donelson,J.E. (1981) Sequence homologies near the C-termini of the variable surface glycoproteins of *Trypanosoma brucei*. *Nature*, **294**, 53–57.
Roberts,J.A. and Tracey-Patte,P. (1975) *Babesi rodhaini* immunoinduction of antigenic variation. *Int. J. Parasitol.*, **5**, 573–576.
Rovis,L. and Dube,D.K. (1981) Studies on the biosynthesis of the variant surface glycoprotein of *Trypanosoma brucei*: sequence of glycosylation. *Mol. Biochem. Parasitol.*, **4**, 77–93.
Seed,J.R. (1964) Antigenic similarity among culture forms of the brucei group of *Trypanosomes. Parasitology*, **54**, 593–596.
Seed,J.R. (1979) The role of immunoglobulins in immunity to *Trypanosoma brucei gambiense*. *Int. J. Parasitol.*, **7**, 55–60.
Steinert,M. and Pays,E. (1985) Genetic control of antigenic variation in trypanosomes. *Br. Med. Bull.*, **41**, 149–155.
Strickler,J.E. and Patton,C.L. (1980) *Trypanosoma brucei brucei*: inhibition of glycosylation of the major variable surface coat glycoprotein by tunicamycin. *Proc. Natl. Acad. Sci. USA*, **77**, 1529–1533.
Strickler,J.E. and Patton,C.L. (1982) *Trypanosoma brucei*: nearest-neighbor analyses on the major variable surface coat glycoprotein—cross-linking pattern with intact cells. *Exp. Parasitol.*, **53**, 117–131.
Tetley,L., Vickerman,K. and Moloo,S.K. (1981) Absence of a surface coat from metacyclic *Trypanosoma vivax*: possible implications for vaccination against vivax trypanosomiasis. *Trans. R. Soc. Trop. Med. Hyg.*, **75**, 409–414.
Turner,M.J. (1984) The biochemistry of variant surface glycoproteins of the African trypanosomes. *Biochem. Soc. Symp.* **49**, 169–181.

Udeinya,I.J., Miller,L.H., McGregor,I.A. and Jensen,J.B. (1983) *Plasmodium falciparum* strain-specific antibody blocks binding of infected erythrocytes to amelanotic myeloma cells. *Nature,* **303**, 429–431.

van der Ploeg,L.H.T. and Cornelissen,A.W.C. (1984) The contribution of chromosomal translocations to antigenic variation in *Trypanosoma brucei*. *Phil. Trans. R. Soc. Lond. Ser. B,* **307**, 13–26.

van der Ploeg,L.H.T., Bernards,A., Rijsewijk,F.A.M. and Borst,P. (1982a) Characterization of the DNA duplication transposition that controls the expression of two genes for variant surface glycoproteins in *Trypanosoma brucei*. *Nucleic Acids Res.,* **10**, 593–609.

van der Ploeg,L.H.T., Lieu,A.Y.C., Michels,P.A.M., De Lange,T., Borst,P., Majumder,H.K., Weber,H., Veeneman,G.H. and Van Boom,J. (1982b) RNA splicing is required to make the messenger RNA for a variant surface antigen in trypanosomes. *Nucleic Acids Res.,* **10**, 3591–3604.

van der Ploeg,L.H.T., Valerio,D., De Lange,T., Bernards,A., Borst,P. and Grosveld,P.G. (1982c) An analysis of cosmid clones of nuclear DNA from *Trypanosoma brucei* shows that the genes for VSGs are clustered in the genome. *Nucleic Acids Res.,* **10**, 5905–5923.

van der Ploeg,L.H.T., Cornelissen,A.W.C., Michels,P.A.M. and Borst,P. (1984a) Chromosome rearrangements in *Trypanosoma brucei*. *Cell,* **39**, 213–221.

van der Ploeg,L.H.T., Schwartz,D.C., Cantor,C.R. and Borst,P. (1984b) Antigenic variation in *Trypanosoma brucei* analyzed by electrophoretic separation of chromosome-sized DNA molecules. *Cell,* **37**, 77–84.

Van Meirvenne,N., Janssens,P.G. and Magnus,E. (1975) Antigenic variation in syringe passaged populations of *Trypanosoma (Trypanazoon) brucei*. Rationalization of the experimental approach. *Ann. Soc. Belge Med. Trop.* **55**, 1–23.

Vickerman,K. (1969) On the surface coat and flagellar adhesion in trypanosomes. *J. Cell Sci.,* **5**, 163–193.

Williams,R.O., Young,J.R. and Majiwa,P.A.O. (1979) Genomic rearrangements correlated with antigenic variation in *Trypanosoma brucei*. *Nature,* **282**, 847–849.

Williams,R.O., Young,J.R. and Majiwa,P.A.O. (1982) Genomic environment of *T. brucei* VSG genes: presence of a mini chromosome. *Nature,* **299**, 417–421.

Wright,K.A. and Hales,H. (1970) Cytochemistry of the pellicle of bloodstream forms of *Trypanosoma (Trypanazoon) brucei*. *J. Parasitol.,* **56**, 671–683.

Young,J.R. (1985) Molecular genetics of antigenic variation in African trypanosomes. Ph.D. Thesis, The University of Cambridge.

CHAPTER 2

Antigenic variation in Hong Kong influenza virus haemagglutinins

J.J. SKEHEL AND D.C. WILEY

National Institute for Medical Research, Mill Hill, London, NW7, UK and Department of Biochemistry and Molecular Biology, Harvard University, 7 Divinity Avenue, Cambridge, MA 02138, USA

The surface membranes of influenza viruses contain two types of virus-specified glycoproteins, haemagglutinins and neuraminidases, each virus containing about 500 haemagglutinin and 100 neuraminidase molecules. The antigenic properties of both glycoproteins differ extensively between strains of virus and are used as the basis of the subtype classification of influenza A viruses. Thirteen distinct haemagglutinins (H_1-H_{13}) and nine neuraminidases (N_1-N_9) have been characterized and viruses have been isolated which contain different combinations of haemagglutinins and neuraminidases. For example, H_3N_2 is the subtype in which the Hong Kong viruses which have infected humans since 1968 are placed; viruses of the H_3N_8 subtype have been frequently isolated from horses and birds, and a virus isolated from quails in 1965 is a member of the $H_{10}N_8$ subtype. Antibodies against either of the virus membrane glycoproteins influence the outcome of infection but antibodies against haemagglutinins directly neutralize virus infectivity (Laver and Kilbourne, 1966). Consequently, viruses with the ability to overcome exisiting immunity and, therefore, with the potential to cause epidemics, contain antigenically novel haemagglutinins. This chapter is specifically concerned with the molecular basis of antigenic variation of haemagglutinins with particular reference to the H_3 subtype.

There have been two influenza pandemics since the first influenza viruses were isolated from humans in 1933 (Smith *et al.*, 1933), the Asian influenza pandemic in 1957 and the Hong Kong pandemic in 1968. Between 1918 and 1957 viruses of the H_1N_1 subtype were prevalent, between 1957 and 1968 viruses of the H_2N_2 subtype, and from 1968 until now viruses of the H_3N_2 subtype. Haemagglutinins of these three subtypes are by definition antigenically distinct and their amino acid sequences are between 25% and 40% homologous [H_1-H_2, 67%; H_2-H_3, 41%; and H_1-H_3, 42% (Winter *et al.*, 1981; Gething *et al.*, 1980; Verhoeyen *et al.*, 1980)]. The abrupt subtype changes in 1957 and 1968 appear to result from the introduction into the human population of influenza viruses from other species and the mechanism of introduction probably involves genetic re-assortment. The ease with which influenza virus genes re-assort during mixed infections both *in vivo* and *in vitro* is well documented (Webster *et al.*, 1971) and is presumed to be a consequence of the segmented nature of the virus genomes, each of which contains eight separate RNA molecules (Palese, 1977). For the H_3N_2 viruses, genome analyses by RNA−RNA hybridization (Scholtissek *et al.*, 1978) indicate that seven of the eight RNAs of the 1968 Hong Kong virus were indistinguishable from the equivalent segments of the genome of the 1967 H_2N_2 virus but that the eighth RNA, encoding the haemagglutinin, was different. Since viruses with haemagglutinins structurally and antigenically similar to that of the 1968 virus are known to infect birds and horses (Waddell *et al.*, 1963; Laver and Webster, 1973) it is generally considered that the 1968 virus was generated by genetic re-assortment in a mixed infection involving one such virus and the 1967 H_2N_2 virus from humans. Subsequent nucleotide sequence analysis of the genes for the haemagglutinins of representative avian and equine viruses have shown (Fang *et al.*,1981; Ward and Dopheide, 1981; Daniels *et al.*, 1985) that the haemagglutinins from the avian virus A/duck/Ukraine isolated in 1963 and the Hong Kong virus are 95% homologous in amino acid sequence whereas those from equine viruses were 91% homologous with the Hong Kong haemagglutinin. An avian origin for the Hong Kong haemagglutinin is therefore favoured.

During interpandemic periods, more gradual changes in the antigenic properties of haemagglutinins occur which are occasionally associated with epidemics of almost the same importance as the initial pandemic. For the Hong Kong viruses, such serious epidemics occurred in 1972 and 1975 and analyses of the structure and amino acid sequences of the haemagglutinins of these viruses and of others isolated each year between 1968 and 1985 have given an understanding of the molecular basis of haemagglutinin antigenicity and of the process of antigenic drift (Both *et al.*, 1983; Skehel *et al.*, 1983).

The haemagglutinin of the Hong Kong virus is a trimer of identical subunits of molecular weight about 220 000 and its three-dimensional structure is known (Wilson *et al.*, 1981). Each subunit contains two polypeptide chains (HA_1 and HA_2) linked by a single disulphide bond. The smaller glycopolypeptide (HA_2) of each subunit contains a hydrophobic region which anchors the molecule in the virus membrane and also forms a prominent α-helix which, in the trimer, interacts with the helices of the two other subunits to form a central fibrous stem. The HA_1 glycopolypeptide chains are mainly involved in the formation of globular domains further removed from the virus membrane which contain receptor binding sites through which the haemagglutinin fulfils its function of attaching virus particles to cells to be infected. The amino acid substitutions detected in haemagglutinins of viruses isolated from epidemics and outbreaks of influenza between 1968 and 1985 are almost exclusively in these distal globular domains (Both *et al.*, 1983; Skehel *et al.*, 1983; Wiley *et al.*, 1981). They appear to cluster in five regions designated sites A−E (*Figure 1*) which overlap. The amino acid residues involved are predominantly at the surface of the molecule which would be directly recognizable by antibodies, except for those residues in sites D which are located in the interfaces between the distal globular domains in the trimer. The assumed antigenic importance of the amino acid substitutions in HA_1 is supported by experiments designed to map antibody binding sites, in which the genes for the haemagglutinins of monoclonal antibody-selected antigenic variants were sequenced (Gerhard *et al.*, 1981; Webster and Lever, 1980; Skehel *et al.*, 1984). The importance of each of the five sites has been confirmed in this way since all of the selected variants, which in the

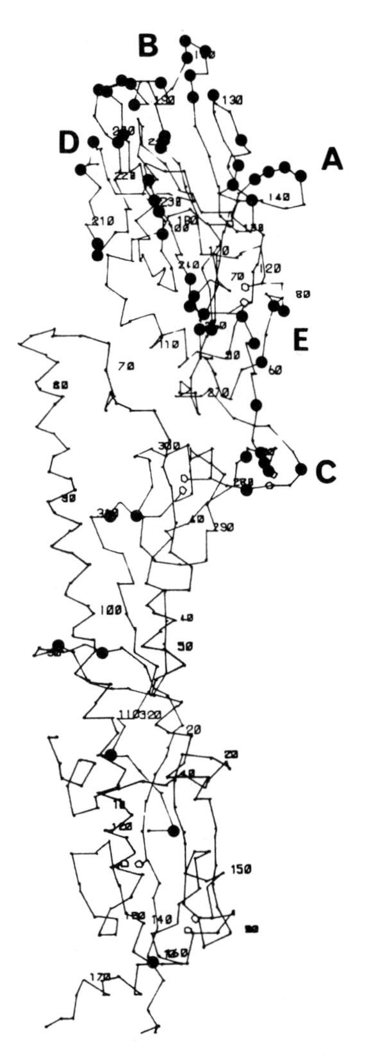

Fig. 1. A schematic diagram of the α-carbon backbone of a haemagglutinin subunit to show the antigenic sites A, B, C, D and E. The locations of a number of the amino acid substitutions clustered within the sites are indicated by the symbols. For details of the structure see Wilson *et al.* (1981).

main contain a single amino acid substitution, are modified in one or other of sites A–E. The identity of the sites of amino acid substitution and antibody binding has been shown clearly in one case by determining the crystal structure of a mutant haemagglutinin containing a single amino acid substitution in site A (Knossow *et al.*, 1984). Since only local changes in structure attributable to the substitution were detected, the antibody used in the selection of the mutant must bind to this region of the molecule and be prevented from binding simply by the addition of the new amino acid side chain. Similar conclusions concerning the distinct sites of antibody binding have been drawn from electron microscopic analyses of antibody–haemagglutinin complexes (Wrigley *et al.*, 1983). The amino acid substitution in the naturally occurring antigenic variants and the monoclonal antibody-selected variants involve changes in side chain charge or size and are frequently identical at given positions in the two classes of variant. In addition, amino acid substitutions have been detected in both classes which lead to additional glycosylation of the haemagglutinin and the antigenic importance of such modifications has been shown experimentally (Skehel *et al.*, 1984). From sequencing studies, the rate of variation in the sequence of the HA_1 glycopolypeptide during the last 15 years has been estimated to be about 0.8% per year.

Overall, these studies indicate that amino acid substitutions in antigenically important regions of the haemagglutinin have accumulated in viruses isolated since 1968 and that viruses containing such modified haemagglutinins have, on a number of occasions, caused epidemics. The experiments with monoclonal antibodies indicate that binding of antibodies to any one of a number of distinct sites on the haemagglutinin neutralizes virus infectivity. It seems reasonable also to conclude, therefore, that significant antigenic drift, which would lead to the generation of variants capable of infecting individuals immune to viruses previously in circulation, should involve amino acid substitutions in each antigenic site. The actual process of selection of such significantly different mutants is not clear but it seems likely that it involves the sequential accumulation of amino acid substitutions which occurs during a number of infections of partially immune individuals. This is suggested by the analysis of the anti-haemagglutinin antibodies in human sera which indicates that people respond differently to infection by generating antibodies with different ranges of specificity (Wang *et al.*, 1986). As a consequence, following a single infection, a significant proportion do not develop complete immunity to the range of antigenic variants in circulation.

To summarize, the two processes of antigenic variation in influenza viruses have different mechanisms. Antigenic drift involves the immune selection of antigenic mutants resulting from point mutations in the genes for haemagglutinin, and occurs continuously in interpandemic periods. It has been assumed here that the selective immune pressure is mainly exerted by antibodies able to neutralize virus infectivity by binding to virus haemagglutinins, but it is also possible that components of the immune system other than B cells may be involved in the selection process. If this is the case, such components would appear to recognize the same surface regions of the haemagglutinin as those identified to be antibody binding sites. Antigenic shifts, on the other hand, which lead to the introduction of viruses with different haemagglutinins into the human population, appear to occur following the re-assortment of virus genes during mixed infections. The recombinant viruses which result have the potential to replicate in humans and are presented to a population without immunity. As a consequence, they spread rapidly to cause pandemics. It is also possible that the impact of these pandemics is influenced by immune cells, and cross-reactive cytotoxic T cells which could fulfil such a role have certainly been characterized (Braciale, 1979). To date, however, the precise specificity of haemagglutinin recognition by such cells has not been determined.

Finally, an additional selective pressure which presumably limits the transfer to humans of recombinant viruses with haemagglutinins optimally adapted for replication in alternative hosts, is imposed by the cellular receptors to which haemagglutinins must bind to initiate infection (Rogers *et al.*, 1985). The different specificities with which the haemagglutinins of viruses from different species recognize sialic acid-containing receptors is reflected in structural differences in their receptor binding sites. Thus, the frequency with which appropriate mutations in these sites occur may dictate the rate of emergence of pandemic influenza viruses.

References

Both,G.W., Sleigh,M.J., Cox,N.J. and Kendal,A.P. (1983) Antigenic drift in influenza virus H3 hemagglutinin from 1968 to 1980: multiple evolutionary pathways and sequential amino acid changes at key antigenic sites. *J. Virol.*, **48**, 52–60.

Braciale,T.J. (1979) Specificity of cytotoxicity T cells directed to influenza virus haemagglutinin. *J. Exp. Med.*, **149**, 856–865.

Daniels,R.S., Skehel,J.J. and Wiley,D.C. (1985) Amino acid sequences of haemagglutinins of influenza viruses of the H3 subtype isolated from horses. *J. Gen. Virol.*, **66**, 457–464.

Fang,R., Min Jou,W., Huylebroeck,D., Devos,R. and Fiers,W. (1981) Complete structure of A/duck/Ukraine/63 influenza haemagglutinin gene: Animal virus as progenitor of human H3 Hong Kong 1968 influenza haemagglutinin. *Cell*, **25**, 315–323.

Gerhard,W., Yewdell,J., Frankell,M.E. and Webster,R. (1981) Antigenic structure of influenza virus haemagglutinin defined by hybridoma antibodies. *Nature*, **290**, 713–717.

Gething,M.-J., Bye,J., Skehel,J.J. and Waterfield,M.D. (1980) Cloning and DNA sequence of double-stranded copies of haemagglutinin genes from H2 and H3 strains elucidates antigenic shift and drift in human influenza virus. *Nature*, **287**, 301–306.

Knossow,M., Daniels,R.S., Douglas,A.R., Skehel,J.J. and Wiley,D.C. (1984) Three-dimensional structure of an antigenic mutant of the influenza virus haemagglutinin. *Nature*, **311**, 678–680.

Laver,W.G. and Kilbourne,E.D. (1966) Identification in a recombinant influenza virus of structural proteins derived from both parents. *Virology*, **30**, 493–501.

Laver,W.G. and Webster,R.G. (1973) Studies on the origin of pandemic influenza. III. Evidence implicating duck and equine influenza viruses as possible progenitors of the Hong Kong strain of human influenza. *Virology*, **51**, 383–391.

Palese,P. (1977) The genes of influenza virus. *Cell*, **10**, 1–10.

Rogers,G.N., Daniels,R.S., Skehel,J.J., Wiley,D.C., Wang,X.-F., Higa,H.H. and Paulson,J.C. (1985) Host-mediated selection of influenza virus receptor variants. *J. Biol. Chem.*, **260**, 7362–7367.

Scholtissek,C., Rhode,W., von Hoyningen,V. and Rott,R. (1978) On the origin of the human influenza virus subtypes H2N2 and H3N2. *Virology*, **87**, 13–20.

Skehel,J.J., Daniels,R.S., Douglas,A.R. and Wiley,D.C. (1983) Antigenic and amino acid sequence variations in the haemagglutinins of type A influenza viruses recently isolated from human subjects. *Bull. W.H.O.*, **61**, 671–676.

Skehel,J.J., Stevens,D.J., Daniels,R.S., Douglas,A.R., Knossow,M., Wilson,I.A. and Wiley,D.C. (1984) A carbohydrate side chain on hemagglutinins of Hong Kong influenza viruses inhibits recognition by a monoclonal antibody. *Proc. Natl. Acad. Sci. USA*, **81**, 1779–1783.

Smith,W., Andrewes,C.H. and Laidlaw,P.O. (1933) A virus obtained from influenza patients. *Lancet*, **ii**, 66–68.

Verhoeyen,M., Fang,R., Min Jou,W., Devos,R., Huylebroeck,D., Saman,E. and Fiers,W. (1980) Antigenic drift between the haemagglutinin of the Hong Kong influenza strains A/Aichi/2/68 and A/Victoria/3/75. *Nature*, **286**, 771–776.

Waddell,G.H., Teigland,M.B. and Sigel,M.M. (1963) A new influenza virus associated with equine respiratory disease. *J. Am. Vet. Med. Assoc.*, **143**, 587–590.

Wang,M.-L., Skehel,J.J. and Wiley,D.C. (1986) Comparative analyses of the specificities of anti-influenza hemagglutinin antibodies in human sera. *J. Virol.*, **57**, 124–127.

Ward,C.W. and Dopheide,T.A. (1981) Evolution of the Hong Kong influenza A subtype. Structural relationship between the haemagglutinin from A/duck/Ukraine 63 (Hav 7) and the Hong Kong (H3) haemagglutinins. *Biochem. J.*, **195**, 337–340.

Webster,R.G. and Laver,W.G. (1980) Determination of the number of non-overlapping antigenic areas on Hong Kong (H3N2) influenza virus HA with monoclonal antibodies and the selction of variants with potential epidemiological significance. *Virology*, **104**, 139–148.

Webster,R.G., Campbell,C.H. and Granoff,A. (1971) The "*in vivo*" production of 'new' influenza viruses. I. Genetic recombination between avian and mammalian influenza viruses. *Virology*, **44**, 317–328.

Wiley,D.C., Wilson,I.A. and Skehel,J.J. (1981) Structural identification of the antibody-binding sites of Hong Kong influenza haemagglutinin and their involvement in antigenic variation. *Nature*, **289**, 373–378.

Wilson,I.A., Skehel,J.J. and Wiley,D.C. (1981) Structure of the haemagglutinin membrane glycoproteiu of influenza virus at 3Å resolution. *Nature,* **289**, 366–373.
Winter,G., Fields,S. and Brownlee,G.G. (1981) Nucleotide sequence of the haemagglutinin of a human influenza virus H1 subtype. *Nature,* **292**, 72–75.
Wrigley,N.G., Brown,E.B., Daniels,R.S., Douglas,A.R., Skehel,J.J. and Wiley,D.C. (1983) Electron microscopy of influenza haemagglutinin-monoclonal antibody complexes. *Virology,* **131**, 308–314.

CHAPTER 3

Antigenic variation in the lentiviruses that cause visna – maedi in sheep and arthritis – encephalitis in goats

OPENDRA NARAYAN[1,3], JANICE E.CLEMENTS[1,2], SUZANNE KENNEDY-STOSKOPF[3] and WALTER ROYAL,III[1]

The Johns Hopkins University, School of Medicine, [1]Departments of Neurology, [2]Molecular Biology and Genetics, and [3]Comparative Medicine, A. Meyer Building, Room 6-181, 600 N.Wolfe Street, Baltimore, MD 21205, USA

Introduction

Lentiviruses comprise an unofficial taxonomic group of viruses that are so named (lenti = slow, L) because they cause diseases characterized by long incubation periods, insiduous onset and slowly progressive clinical courses (Sigurdsson, 1954). The viruses include the etiologic agents of equine infectious anaemia (Cheevers and McGuire, 1985), visna – maedi of sheep (Haase, 1975), caprine arthritis – encephalitis (CAE) of goats (Crawford *et al.*, 1980) and acquired immune deficiency syndrome (AIDS) in humans (Popovic *et al.*, 1984). The molecular studies linking these viruses together have been reported by Gonda *et al.* (1985, 1986). The agents are retroviruses—enveloped RNA viruses that have RNA-dependent DNA polymerase and replicate by means of a proviral DNA template (Lin and Thormar, 1970; Haase and Varmus, 1973). However, unlike most retroviruses, they are not oncogenic.

The lentivirus genome is approximately 10 kb which is divided into three structural genes; the *gag* which codes for the internal core proteins of the virions, the *pol* which codes for the reverse transcriptase, endonuclease/integrase enzymes and the *env* which encodes the envelope glycoprotein(s) of the virus and the proteins that induce neutralizing antibodies (Scott *et al.*, 1979). The proviral DNA has flanking long terminal repeats (LTR) which contain the promoter and enhancer elements that control viral RNA transcription (Hess *et al.*, 1985). In addition, small open reading frames located between the *pol* and *env* genes in the proviral DNA are thought to encode non-structural proteins that act in *trans* on the enhancer to modulate viral gene expression (Sonigo *et al.*, 1985; Wain-Hobson *et al.*, 1985). The net result of these regulatory processes is an extreme of two types of viral replication: a highly productive type in cell cultures of the host (Clements *et al.*, 1978); Haase *et al.*, 1982) and an unusual restricted type in tissue cells of the infected animal (Narayan *et al.*, 1977a; Haase *et al.*, 1977). Productive virus replication in cell culture results in virus-induced cytopathology characterized by multinucleated giant cell formation leading to lysis (Sigurdsson *et al.*, 1960;

Narayan *et al.*, 1980). Lytic plaques develop when inoculated cultures are overlaid with agarose. This has been used as both a measure of infectivity in tissue samples from infected animals and as a mechanism for obtaining pure virus populations from a single infectious unit. In contrast, infection in the animal results in 'slow virus' replication with minimal production of cell free virus (Narayan *et al.*, 1977a; Cork and Narayan, 1980). The agent does not cause any direct deleterious effects *in vivo*.

Disease in sheep and goats results from immunopathological processes that involve accumulations of massive numbers of mononuclear inflammatory cells in specific tissues and organs, with resulting disruption of physiological functions (Nathanson *et al.*, 1976; Narayan and Cork, 1985). Visna is characterized by slow onset of progressive paralysis resulting from an encephalitic process (Sigurdsson *et al.*, 1957; Sigurdsson and Palsson, 1958) and maedi, by respiratory disease resulting from chronic interstitial pneumonia (Gudnadottir and Palsson, 1967; Georgsson and Palsson, 1971). The diseases in goats are related to similar chronic inflammatory responses in the brain, joints and mammary gland (Narayan and Cork, 1985).

Visna–maedi and CAE viruses have a tropism for cells of the monocyte–macrophage lineage and the restriction of virus replication in the animal is associated with maturational processes in these cells (Narayan *et al.*, 1982, 1983; Gendelman *et al.*, 1985). The agent is latent in promonocytes in the bone marrow and in monocytes in peripheral blood. Maturation of infected monocytes into macrophages in selected tissues is accompanied by increased levels of viral RNA transcription and translation of viral RNA (Gendelman *et al.*, 1985, 1986). However, virion assembly is still restricted. In the mature macrophages this type of restriction is associated with an interferon that is induced during interaction between lymphocytes and infected macrophages (Narayan *et al.*, 1985). The interferon restricts virus replication in macrophages and this appears to occur at the level of transcription (Kennedy *et al.*, 1985). Thus, in the animal, maintenance of the virus in cells of the monocyte–macrophage lineage provides a unique mechanism for virus persistence and restricted replication and probably forms the basis for lifelong infection.

Virus isolation from infected sheep

Inoculation of sheep with visna–maedi virus results in lifelong infection as described above. The virus is not present in plasma and only small amounts are found in tissue homogenates (Narayan *et al.*, 1977b). However, virus could be obtained from animals at any time by co-cultivating leukocytes from blood with sheep fibroblast cell cultures. Since sheep cell cultures are highly permissive for virus replication, they provide a mechanism for amplifying virus in latently-infected leukocytes. Using this amplification system to advantage in infectious centre assays, we showed that the virus is associated exclusively with the monocyte fraction in the blood (Narayan *et al.*, 1982). However, less than 1% of these cells are infectious. This emphasizes the level of restriction of virus replication in the animal. Routine virus isolation from the animal therefore requires co-cultivation of leukocytes from approximately 10–50 ml of blood with the fibroblast cell cultures (Narayan *et al.*, 1978).

Antibody responses of infected sheep and identification of antigenic variants

Sheep inoculated with visna virus produce neutralizing antibodies several weeks after inoculation. These antibodies in plasma co-exist indefinitely with infected monocytes in the blood. However, early studies on persistent infection in sheep had shown that virus isolated from the animals late in the infection could not be neutralized by serum antibodies (Gudnadottir, 1974). This suggested that the virus may mutate during persistent infection and that non-neutralizable mutants are selected by the antibodies. In order to define the virus mutation/antibody selection hypothesis more clearly, we inoculated sheep with virus grown in tissue culture from a single plaque. Serum from blood and virus from leukocytes were collected at various intervals during a 3 year period. Tests for neutralization of viruses isolated at multiple intervals with serum collected early (1–6 months) and late (2–3 years) after inoculation were performed. These results showed that in all cases the same dilution of early immune serum that neutralized the virus used for inoculations also neutralized viruses obtained from the peripheral blood leukocytes (PBL) early after infection. These viruses were thus indistinguishable antigenically. In some animals, virus obtained late in infection was also similar anti-

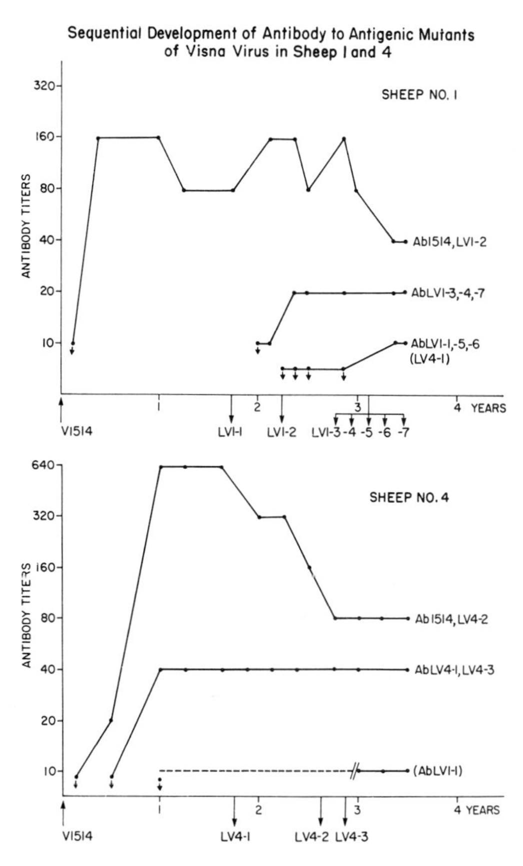

Fig. 1. Sheep no. 1 (**a**) and no. 4 (**b**) were inoculated intracerebrally with a plaque-purified stock of virus 1514 (arrow pointing upwards). The identification and time of recovery of viruses (LV) from peripheral blood leukocytes of these animals is indicated by arrows pointing downwards. Line graphs indicate the temporal development and levels of neutralizing antibody (Ab) in these animals to the strain of virus inoculated and also to the viruses isolated from blood. Antibody titres are reported as the last doubling dilution of serum which neutralized 100 TCD_{50} of virus. Reprinted with permission from Narayan *et al.* (1978).

genically to the inoculum virus, despite the fact that the sheep had produced neutralizing antibodies several months before (Narayan *et al.*, 1978; Lutley *et al.*, 1983; Thormar *et al.*, 1983). This showed that virus phenotypes could remain unaltered in persistently-infected sheep for long periods, in the presence of specific neutralizing antibodies. However, in other animals, viruses obtained at late intervals were not neutralized by antibodies in the early sera (Narayan *et al.*, 1977a, 1978; Lutley *et al.*, 1983). This confirmed earlier results on antigenic drift of visna virus, supporting the concept that mutation of the virus could occur during persistent infection and that non-neutralizable mutant viruses are selected by the antibodies.

In order to explore further the role of mutant viruses in persistent infection, two animals (No. 1 and No. 4), from which mutant viruses had been obtained, were selected for further study (Narayan *et al.*, 1978). One question that arose was whether virus of the parental phenotype had been replaced by mutants during persistent infection. Results of this are shown in *Figure 1*. Between 1.5 and 3 years after inoculation of these animals, PBLs from blood were suspended in medium and added to flasks containing indicator fibroblast cell cultures, PBL from a minimum of 10 ml of blood being required for virus isolation. Virus from each culture was then examined for neutralization by early serum from the same animal. This experiment showed that in sheep no. 1, one of the seven isolates (LV1-2) was indistinguishable from the parental virus used for inoculation, i.e. the same dilution of serum which neutralized the inoculum virus (V1514) also neutralized LV1-2. The other viruses showed reduced or no susceptibility to neutralization by the serum but the animal gradually acquired antibodies to all the viruses late in the infection. A similar sequence of events occurred in sheep no. 4. One of the viruses isolated from PBL was antigenically indistinguishable from V1514, and two were more distant. It was of interest that these antibodies did not cause elimination of any of the viruses. Furthermore, virus of the parental phenotype was not replaced by antigenic mutants in either sheep. These data suggested therefore that persistence of visna virus in the immune environment does not require mutation, agreeing with reports by Lutley *et al.* (1983) and Thormar *et al.* (1983). Whether antigenic mutation conferred any other advantages to the virus remains to be determined.

Molecular basis of antigenic variation

We performed RNase T1 fingerprinting and mapping of the genomes of parental and mutant viruses in order to determine the molecular basis of the altered antigenicity (Clements *et al.*, 1980). Purified RNA from each of the seven viruses from sheep no. 1 described above was cleaved with RNase T1 and the resulting oligonucleotides labelled with ^{32}P and separated by two-dimensional polyacrylamide gel electrophoresis. The migration pattern of the oligonucleotides from each virus was visualized in autoradiograms, producing a fingerprint. Comparison of the autoradiograms of the viruses showed a very similar pattern, indicating that there were no major genetic differences between parental and mutant viruses. However, there were changes in a small subset of oligonucleotides and unique changes among these appeared to be specific for each virus (Clements *et al.*, 1982). Further, analysis of sized, poly(A)-tailed, subgenomic fragments of the viral RNA showed that nearly all of these unique oligonucleotides were located in the 3′ end of the genome (i.e. the *env* gene).

Table 1. Comparison of oligonucleotides and neutralization susceptibility of parental and mutant visna viruses derived in sheep and in culture.

Virus	*Oligonucleotides present in the env gene*							*Neutralizing titre*
Parental virus	–	–	–	16, 45, 29, 42,	21	–	24	160
LV1-2	–	–	–	16, 45, 29, 42,	104	–	–	160
LV1-7	–	–	–	16, 45, 29, 42,	104	–	–	20
LV1-4	–	–	–	16, 45, 29, 42,	104	–	–	20
LV1-3	–	[103	]	16, 45, 29, 42,	104	–	–	20
LV1-6	–	[–,	103]	16, 45, 29, 42,	104	–	–	<10
LV1-5	–	[103A,	–]	16, 45, 29, 42,	104	–	–	<10
LV1-1	102	[103A,	103B]	16, 45, 29, 42,	104,	105,	106	<10
Early serum mutant	–	–	–	16, 45, 29, 42,	104,	105,	24	<10

Comparison of the unique oligonucleotides of the parental virus with mutant viruses showed that the mutations range from two in LV1-2 to several in LV1-1, the non-neutralizable virus. Furthermore, the mutations were cumulative (*Table 1*). Since the mutations were located in the *env* gene, (the gene that codes for the neutralization antigens), it is highly probable that these changes were responsible for the neutralization (or lack of) of the virus by the antibodies in early sera. As shown in *Table 1*, this is supported by the strong correlation between the number of accumulated mutations and resistance to neutralization. Also, since the mutations were cumulative, the viruses must have evolved sequentially, one from the other, during the restricted replication process *in vivo*. The fact that all the viruses were circulating in PBL simultaneously is consistent with the inability of the host defence mechanisms to eliminate them. Whether any of the mutants was dominant at the time of isolation was not determined because too few isolations were performed.

Selection of antigenic mutants by early immune sera

The slow development of antigenic mutants in the sheep was probably a direct reflection of the restricted rate of virus replication in the animal. In order to establish a direct relationship between mutation of the virus and selecting antibodies, we attempted to duplicate events in the animal in the cell culture system, in which virus replication is much faster. Initial experiments were conducted to determine whether virus stocks already had pre-existing antigenic mutant viruses that could be recognized by antibodies in 'early' sera. Plaque-purified virus was expanded in cell culture and supernatant fluid containing 1×10^7 p.f.u. of virus was mixed with an equal volume of 1 in 10 dilution of serum and incubated at 37°C. *Figure 2* shows the gradual loss of infectivity caused by neutralization of the virus. After 6 h at 37°C and overnight at 4°C, the virus–serum mixture was examined for residual infectivity. The mixture was diluted and the entire suspension inoculated into cell cultures in order to rescue and amplify virus that was not neutralized. After 2 h of incubation at 37°C to allow for adsorption and penetration, the cultures were rinsed to get rid of the antibody, overlaid with agarose and incubated at 37°C. Thirty one plaques developed in these cultures. Virus obtained from each plaque was expanded by one passage in culture and each suspension tested with

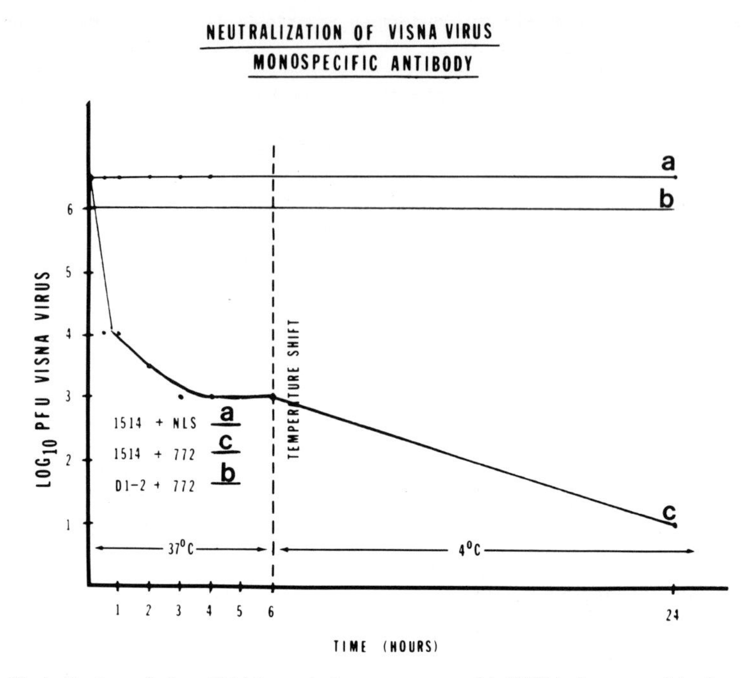

Fig. 2. Neutralization of virus 1514 by early immune serum. (**a**) V1514 plus normal lamb serum (NLS) (**b**) Heterologous visna virus, D1-2 plus immune serum from sheep no. 772. (**c**) V1514 plus immune serum from sheep no. 772. Note lack of neutralization in **a** and **b** and progressive loss of infectivity in **c**. Virus that survived after 24 h incubation in **c** was then examined for neutralization by the same serum.

the same 'early' serum for neutralization. This experiment showed that virus in each suspension was neutralized by the antibodies. The residual infectivity from the original virus–antibody mixture thus represented the 'non-neutralizable tail' that is found typically in all neutralization reactions and did not represent antigenic mutants, as defined by the antibodies in the serum. Thus, the virus stock contained less than one antigenic mutant in 1×10^7 p.f.u. of the virus.

Since mutant viruses apparently developed during virus replication in the animal we asked whether mutants would develop also in cell cultures if the latter were inoculated first with virus and then treated with the antisera. A single plaque of visna virus was suspended in 1 ml (the titre was approximately 1×10^3 p.f.u. ml^{-1}) and two cultures were inoculated with 0.5 ml of the suspension. After incubation to allow for virus adsorption and penetration, one flask was replenished with medium containing normal serum and the other with early immune serum containing neutralizing antibodies. *Figure 3* shows that the virus replicated promptly in the culture with normal serum and killed the cells within 10 days. In contrast, both virus replication and development of cytopathic effects (CPE) were delayed in infected cultures treated with the antiserum. Further, whereas virus from the control culture was neutralized by early antibodies, virus from the culture treated with the early antibodies was not neutralized by this serum (Dubois-Dalcq *et al.*, 1979). It is unlikely that this mutant was present in the original plaque suspension containing 1×10^3 p.f.u. because such a mutant could not be detected in the suspension that contained 1×10^7 p.f.u. of virus. Furthermore, subsequent

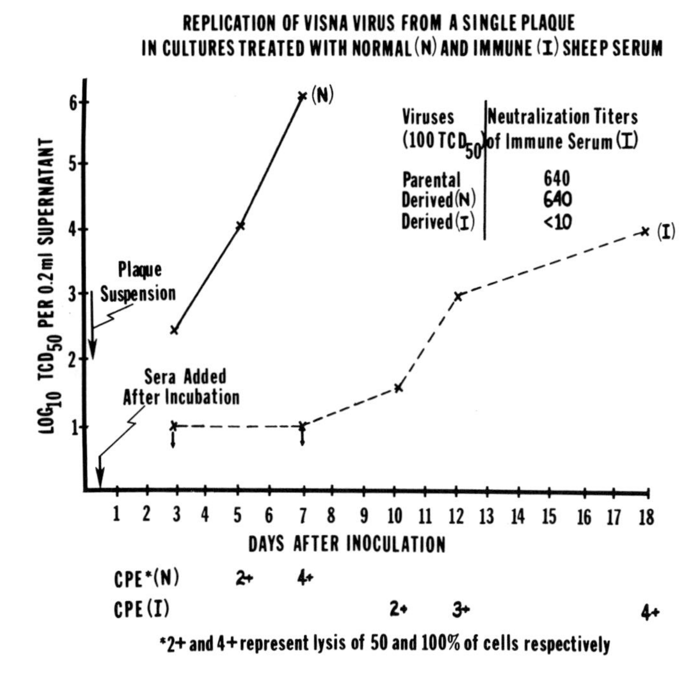

Fig. 3. Growth curve of visna virus in two flasks of sheep choroid plexus cells inoculated with equal portions of a plaque suspension of virus 1514. Cells in one flask (N) were given maintenance medium containing 2% normal lamb serum and the other (I) 2% immune serum (no. 4 sheep). Progression of CPE with time is indicated under the graph. One hundred TCD_{50} of parental virus from which the plaque was derived and virus from the final supernatant fluid for flasks N and I were tested for neutralization by doubling dilutions of the immune serum in a quantal assay. Parental virus and virus from N were neutralized at a dilution of 1/640. Virus from I was not neutralized at a dilution of 1/10. Reprinted with permission from Dubois-Dalcq *et al.* (1979).

experiments showed that mutant viruses could be obtained predictably in every cell culture that was infected with visna virus and treated with the antibodies (e.g. early serum from sheep no. 1 and 4) (Narayan *et al.*, 1978, 1981). The mutant viruses were antigenically stable and replicated as efficiently as the parental virus in fibroblast cell cultures. Fingerprint analysis of one of these mutants showed that it had a subset of the mutations found in LV1-1, the non-neutralizable mutant isolated from the sheep (*Table 1*). This proved that mutation occurred during virus replication and the speed of development of the mutant correlated with the rate of the replication process—slow in the animal and fast in cell culture. The strong selection pressure of antibody in cell culture was attributed to the all-or-none permissive type of virus replication in sheep fibroblasts and the limited number of target cells in a culture dish. In contrast, the lack of selection pressure of the same antibodies in the animal was attributed to the unique restricted type of virus replication in the monocyte–macrophage cell type, lack of virus killing of target cells (i.e. proviral DNA of all viruses is preserved) and the infinite number of potential target cells in tissues (i.e. there was no competition among the viruses for target cells).

Since LV1-1 was not neutralized by early antibodies to parental virus 1514 but appeared to be derived by mutation of the latter agent, we sought to determine the specificities of the antibody response to this virus after inoculation into sheep. The virus was plaque purified and new stocks containing approximately 5 × 10^6 p.f.u. were inoculated into two sheep. Both animals produced neutralizing antibodies to this virus within 6 weeks of inoculation. Surprisingly however, these antibodies neutralized parental virus 1514 at the same dilutions as those that neutralized LV1-1 (Narayan *et al.*, 1978). Thus, LV1-1 contained the antigenic determinants of parental virus 1514.

Early immune sera had a narrow range of neutralizing antibodies but it is evident that the spectrum included antibodies to the neutralization epitopes of not only the virus used for infection of the animal but also the parent from which it mutated. The restriction in neutralization capacity of these sera thus was applicable mainly to the future mutants of the particular virus. The sequence of mutation and selection from the reference strain K1010 (Petursson *et al.*, 1976) may be represented as follows:

K1010 → virus 1514 → LV1-1 → mutant of LV1-1

Antibodies induced by virus 1514 neutralized V1514 and the parental strain from which it was derived (K1010) (Narayan *et al.*, 1981) but not LV1-1 or the mutant of LV1-1. Similarly, antibodies induced by LV1-1 neutralized ancestral viruses K1010 and virus 1514 but not the mutant of LV1-1. Hence progressive mutation had occurred.

Switching over to the cell culture system, we inoculated several dishes of fibroblasts with V1514 and others with LV1-1. After virus adsorption, both sets of cultures were replenished with medium containing antiserum to LV1-1. After 1 month of observations, no mutants had developed in cultures receiving V1514 plus anti-LV1-1. However, mutants were readily obtained in cultures that were inoculated with LV1-1 plus anti-LV1-1. These latter early serum mutants of LV1-1 were not neutralized either by antibodies to LV1-1 or by antibodies induced by parental virus 1514. Thus, it appeared that visna virus mutated progressively under antibody selection pressure and the antigenic profiles of the mutants reflected the specificities of the selecting antibodies (Narayan *et al.*, 1981).

It was surprising that although the early immune sera neutralized ancestral viruses, they did not easily select mutants from cultures infected with these viruses (as seen in the cultures infected with virus 1514 and treated with anti-LV1-1). Since LV1-1 had accumulated multiple mutations in the *env* gene of the parental virus, antibodies of LV1-1 would be unlikely to select further mutants of 1514 virus. This is analogous to selection of mutants by hyperimmune serum described below.

Selection of mutants by late immune sera

Using the same basic scheme of infection of cell cultures followed by treatment with antibodies as described above, we next asked whether the late immune serum would be as efficient in selecting mutant viruses as the early immune serum. The design and results of these experiments are illustrated in *Table 2*. Ten inoculated cultures were treated with the early serum and 10 with the late serum. This experiment showed that all 10 of the cultures treated with early serum produced mutant viruses within 3 weeks. In contrast, only three of the cultures treated with late serum developed infectious virus mutants (heralded by the development of CPE) and this required up to 6 months of incubation (Narayan *et al.*, 1981). Similar results were obtained with early and late sera from other sheep infected with visna virus. Thus, antigenic mutants developed

Table 2. Efficiency of different sera in selection of mutant viruses.

Test	*Early serum*	*Late serum*
Selection of mutants in virus suspension	Negative	ND
Selection of mutants during virus replication	10/10	3/10
Neutralization of early mutants (12)	0/12	12/12
Neutralization of late mutants (3)	0/3	Poor to negative

faster and more predictably when virus was grown in the presence of early serum than in late serum. Extrapolation of these data to the infected animal would therefore imply that mutation and selection of mutant viruses would occur more efficiently during the early phase of infection than the late phase. The similarity of the fingerprint of LV1-1 to that of the early serum mutant V1514 supports this idea, since neither virus was neutralized by early serum but both were neutralizable by late serum.

Antigenicity of mutant viruses (Table 2)

Mutant viruses selected by antibodies in early serum (early serum mutants) and those selected by antibodies in late serum (late serum mutants) were compared in neutralization tests with both types of sera (Narayan *et al.*, 1981). None of 12 early serum mutants or three late serum mutants were neutralized by early serum from infected sheep. In contrast, all early serum mutants were neutralized by the late serum. These results suggested that the early sera had selected specific mutants with a limited number of changes that allowed escape from neutralization by antibodies in the early sera. Nevertheless the mutants were recognized by antibodies in the late serum. Thus, the late serum presumably contained a wider spectrum of antibodies than early serum and recognized the changes in the early mutants and neutralized their infectivity. The failure of the late serum to neutralize late mutants was attributed to further mutations of the virus, to the point where they were no longer recognizable by these antibodies. Since the late sera neutralized early mutants, we speculated that the late mutants had all the changes present in early mutants plus additional changes. Support for this came from the fingerprint studies of the mutant viruses described above (*Table 1*). This showed that LV1-2 virus with only two mutations still maintained neutralization characteristics of the parental phenotype. In contrast, LV1-1, a virus which had the mutations of LV1-2 plus further mutations, was not neutralized by antibodies in the early serum. This virus was only neutralized by antibodies in the late serum (see *Figure 1*). The fact that LV1-1 is biologically and genetically similar to the early serum mutants of V1514 suggested similar mechanisms for their derivation, i.e. selection by early serum. The equivalent of late serum mutants was not found in sheep and would probably be very rare. We have not fingerprinted these late mutants as yet but we anticipate many more mutations in these viruses than those in the early mutants (or LV1-1). Such progressive mutation provides a mechanism for the wide genetic variability of these viruses.

Since visna virus does not have an unusually high spontaneous mutation rate as evidenced by the low frequency of antigenic mutants in stock preparations, the development of such mutants under antibody pressure in culture must have been due to a very narrow range of neutralizing antibodies in early sera. This is unusual for a polyclonal, post-infectious immune response but is nevertheless typical for sheep antibody responses to visna virus. Such a narrow clonal response was overcome later in the infection with development of a wide antibody spectrum more typical of polyclonal neutralizing antibodies. However, the mechanism of formation of these antibodies is not understood. This may be fundamental to the development of antigenic variant viruses. Data in *Figure 1* had suggested that the antibodies may have developed in response to new antigenic determinants in mutant viruses. In contrast, studies from another sheep (in the same study, Narayan *et al.*, 1978), from which no mutant viruses had been obtained, had showed that this animal also produced antibodies of a widening spectrum during 2 years of persistent infection. This led to an alternative hypothesis that the virus may have a specific number of neutralization determinants and that antibodies are made in sequence to each determinant, over a long period of time. To address this question, plaque-purified visna virus was disrupted with detergent and inoculated into a sheep, first in an emulsion with Freund's complete adjuvant and later in emulsions of incomplete adjuvant. Examination of serum from this animal after several injections showed that it had developed antibodies that neutralized both early and late mutants of visna virus (Narayan *et al.*, 1981). This supported the second hypothesis that all epitopes in both early and late mutant viruses were present in a single strain of virus. Thus, mutation probably results in rearrangement of epitopes already in existence in the virion rather than coding for new epitopes. This is shown in diagrammatic form in Figure 4.

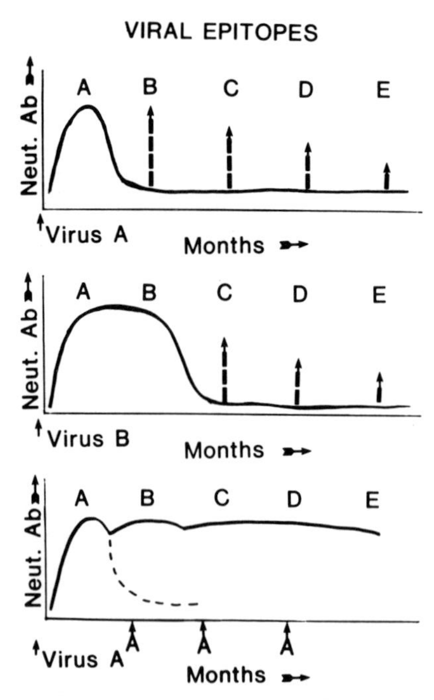

Fig. 4. Diagram showing putative neutralizing antibody responses of sheep to the multi-epitopic visna virus. In the top row, sheep produce antibodies to the first epitope and these antibodies then select the viruses with the second epitope. In the middle row, antibody production to the parental virus in addition to the mutant virus is indicated. In the bottom row, when virus in (A) is used for hyperimmunization of a single sheep, antibodies are produced to all the viruses.

Antigenic variation in caprine arthritis–encephalitis virus

The association of a retrovirus with arthritis and encephalitis in goats was first reported in 1980 (Crawford *et al.*, 1980). Unlike visna–maedi disease, which is sporadic in the United States, the CAE virus infection is epidemic among dairy goats in the U.S. and Western Europe and approximately 10% of infected adult goats develop arthritis (Crawford and Adams, 1981; Adams *et al.*, 1984). CAE and visna–maedi viruses are genetically distinct (Robertson and Cheevers, 1984) but share extensive homology both in genetic sequences and in antigenic determinants, as demonstrated in binding tests such as enzyme-linked immunosorbent assay (ELISA) and immunoprecipitation (Pyper *et al.*, 1984). Further, both viruses have a similar type of pathogenesis in their respective hosts. These include tropism for cells of the monocyte–macrophage lineage, slow restricted replication, persistent infection and immunopathological mechanisms in disease.

Despite the homology between the two viruses (Pyper *et al.*, 1984), antibodies obtained by hyperimmunization of sheep or goats with visna virus do not neutralize the caprine agent (Narayan *et al.*, 1984). A further difference between the two viruses is that CAE virus only rarely induces neutralizing antibodies. Lack of such antibodies could not be overcome by hyperimmunization of goats with purified virus, inactivated virus, virus disrupted with detergents or virus-infected cells (Klevjer-Anderson and McGuire, 1982; Narayan *et al.*, 1984). Whereas immunized goats produced binding antibodies to all virus proteins including the envelope glycoprotein (Johnson *et al.*, 1983)

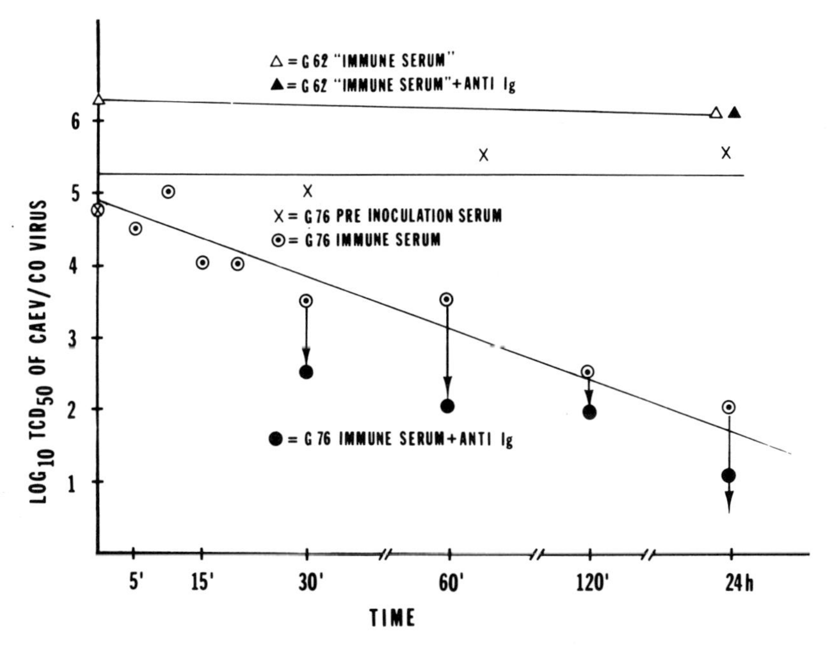

Fig. 5. Neutralization of caprine arthritis virus. Suspensions of CAEV/CO containing 5 × 10^6 TCD_{50} were mixed with serum from goat no. 62 which had been hyperimmunized with visna virus 1514 and CAEV/CO; and from goat no. 76 which had been immunized with CAEV/CO plus *Myobacterium tuberculosis*. Neither G62 serum (△) nor G62 serum supplemented later with anti-goat Ig (▲) caused any reduction in infectivity. In contrast, G76 serum (○) neutralized the virus and the neutralization was enchanced when the serum was supplemented with anti-Ig (●). Reprinted with permission from Narayan *et al.* (1984).

these antibodies did not neutralize infectivity. This lack of neutralizing capacity had prompted the question whether CAE virus even contained neutralization epitopes. However, recent studies have shown that these viruses do have such epitopes because neutralizing antibodies were induced in some goats when virus strain CAEV/CO was administered in conjunction with large amounts of inactivated *Mycobacterium tuberculosis* (Figure 5). Having obtained these neutralizing antibodies, we then commenced experiments to determine whether the virus undergoes antigenic mutation as seen in visna virus.

Initial studies showed that the antibodies (IgG1 class) neutralized only the strain of virus that was used for immunization (CAEV/CO strain). The antibodies did not neutralize any strains of visna virus or field strains of CAE virus. This included CAE viruses obtained from goats from the same herd from which CAEV/CO had been derived, thus providing an initial indication that CAE viruses in nature were antigenically diverse.

Since CAE viruses do not normally induce neutralizing antibodies in infected goats we could not determine whether multiple antigenic types of viruses were circulating in nature in different animals or whether there were antigenic differences among viruses obtained from the same animal. We therefore undertook studies to determine whether antigenic mutants of CAEV/CO, as defined by our neutralizing antibodies, could be detected in a stock preparation of this virus. Using the protocol applied to visna virus (described above), we mixed a suspension containing 1×10^6 p.f.u. with the immune serum, incubated at 37°C for 2 h, and then examined it for residual infectivity. A non-neutralized 'tail' of virus was obtained. However, expansion of progeny from these plaques showed that virus from many of these suspensions could not be neutralized by the antibodies. This result was different from that obtained with visna virus and showed that CAE virus had an antigenic mutation rate greater than 1 in 1×10^6. In order to determine the frequency of this mutation, dilutions of the virus stock containing 10^2, 10^3, 10^4 and 10^5 p.f.u. were mixed with the antiserum, incubated for 2 h at 37°C and inoculated onto indicator cell cultures. No plaques were obtained in cultures inoculated with 10^3 p.f.u. or less. However, cultures inoculated with 1×10^4 p.f.u. or higher all developed plaques. Progeny of four of these plaques was subcultured and examined for neutralization. All were resistant. This showed that the virus had an antigenic mutation rate of 1 in 10^4. More importantly, these mutants had developed without the selection process required for visna virus.

Discussion

These experiments have shown that antigenic mutation is common among ovine and caprine lentiviruses although the mechanisms of development of such mutants appeared to be different between viruses of the two families. The studies with polyclonal antibodies to visna virus suggest that the viral genome encodes a specific number of neutralization epitopes, and that point mutations in the *env* gene result in differential exposure and/or rearrangement of these epitopes rather than synthesis of unique epitopes. Persistently-infected sheep produce antibodies to certain of the epitopes in the inoculum virus and these antibodies may or may not select antigenic variant viruses that have mutations coding for new epitopic arrangements. Mutations are continuous and cumulative. The number of mutations correlates directly with the antigenic disparity between parental and mutant viruses: the greater the number of point mutations, the greater the difference in antigenicity between the parental and the mutant viruses. The antibody spectrum in serum of the persistently infected animal increases with time and this may be due to the eventual production of antibodies to all of the epitopes in the virion. The development of antigenic mutant viruses in only some sheep may be a reflection of differences in the initial neutralizing antibody response produced early in the infection. Possibly, in some sheep, sera may have a very narrow spectrum of antibodies and these then select for viruses with rearranged antigenic mosaics. In contrast, other sheep may produce antibodies initially to several epitopes (resembling the antibody repertoire of late sera) and thus the selection of mutants occurs less frequently or not at all.

As suggested by Lutley *et al.* (1983) and Thormar *et al.* (1983) antigenic mutation of visna virus may not be essential for the survival of the virus in the persistently-infected sheep. The ability of the virus to integrate proviral DNA into host DNA and to infect cells of the monocyte – macrophage series already endow the agent with indefinite survival mechanisms because: (i) the immune system cannot eliminate a latent virus genome, and (ii) the target cell is the macrophage which is one of the major cell types in the defence system charged with elimination of foreign pathogens. Infection in this cell may therefore preclude elimination of the agent. Since visna mutants are replication-competent and virulent, we had surmised in earlier studies that development of mutants late in the infection had the potential for mounting sequential waves of replication and thus precipitating episodes of disease. Although this is possible, the idea is less compelling because of recent unpublished data in our laboratory on the kinetics of interaction between virus and neutralizing antibodies and between the virus and the cell. These studies showed that the rate of virus neutralization was much slower than the process of infection of the cell (Kennedy-Stoskopf and Narayan, 1986). This suggests that virus could spread from cell to cell before the virus could be neturalized, and provides a mechanism for the known ability of visna virus to replicate indefinitely, irrespective of the neutralizing antibodies in plasma.

Antigenic mutation is of less importance for persistence of CAE virus than of visna virus. The fact that goats rarely develop neutralizing antibodies removes the necessity for the virus to adopt the strategies of visna virus in order to evade the immune responses. The low immunogenicity and/or the covert nature of the neutralization epitopes themselves constitute a highly effective mechanism for eluding humoral immune responses of the host. The facts nevertheless speak for themselves: this virus has an extraordinarily high rate of antigenic mutation. This is reflected also by the great genetic variability seen in restriction maps of DNA derived from different field isolates (Yaniv *et al.*, 1985) However, as noted for visna virus, the significance of the mutations in CAE virus is not readily apparent. It is possible that the mutations in these viruses, with attendant biological changes, are epi-phenomena, resulting from errors of reverse RNA transcription by the mistake-prone viral polymerase gene. However, since viruses do not often have luxury functions, another possibility is that antigenic variation may be important for the continuous evolution of the virus family.

As pathogens, these agents present obvious problems for potential control by immunization. Antigenic variation of the virus during persistent infection and antigenic variability of field viruses are one major concern. When this is compounded with the problems of immune responses that extend from antibodies with low affinity and narrow neutralizing spectra to no neutralizing capacity then the concept of protective immunization becomes untenable.

Acknowledgements

These studies span several years and dedicated work by several individuals. The expert technical assistance by Janis Chase, Martha Sommerville, George Fleming and Darlene Sheffer are gratefully acknowledged. We thank Linda Kelly for preparing the manuscript. The studies have been supported by grants NS12127, NS15721, NS16145, RR00130 from the NIH and a gift fund from the Hamilton Roddis Foundation. S.K. was supported by a Special Emphasis Career Award RR00017 from the NIH.

References

Adams,D.S., Oliver,R.E., Ameghino,E., DeMartini,J.C., Verwoerd,D.W., Houwers,D.J., Waghela,S., Gorham,J.R., Hyllseth,B., Dawson,M., Trigo,F.J. and McGuire,T.C. (1984) Global survey of serological evidence of caprine arthritis encephalitis virus infection. *Vet. Rec.*, **115**, 493–495.

Cheevers,W.P. and McGuire,T.C. (1985) Equine infectious anemia virus: Immunopathogenesis and persistence. *Rev. Infect. Dis.*, **7**, 83–88.

Clements,J.E., Narayan,O., Griffin,D.E. and Johnson,R.T. (1978) The synthesis and structure of infectious visna virus DNA. *Virology*, **93**, 377–386.

Clements,J.E., Pederson,F.S., Narayan,O. and Hazeltine,W.S. (1980) Genomic changes associated with antigenic variation of visna virus during persistent infection. *Proc. Natl. Acad. Sci. USA*, **77**, 4454–4458.

Clements,J.E., D'Aontion,N. and Narayan,O (1982) Genomic changes associated with antigenic variation of visna virus. II. Common nucleotide sequence changes detected in variants from independent isolations. *J. Mol. Biol.*, **158**, 415–434.

Cork,L.C. and Narayan,O (1980) The pathogenesis of viral leukoencephalomyelitis–arthritis of goats. I. Persistent viral infection with progressive pathological changes. *Lab. Invest.* **42**, 596–602.

Crawford,T.B. and Adams,D.S. (1981) Caprine arthritis–encephalitis. Clinical features and presence of antibody in selected goat populations. *J. Am. Vet. Med. Assoc.*, **178**, 713–719.

Crawford.T.B., Adams,D.S., Cheevers,W.P. and Cork,L.C. (1980) Chronic arthritis in goats caused by a retrovirus. *Science*, **207**, 997–999.

Dubois-Dalcq,M., Narayan,O and Griffin,D.E. (1979) Cell surface changes associated with mutation of visna virus in antibody-treated cell cultures. *Virology*, **92**, 353–366.

Gendelman,H.E., Narayan,O., Molineaux,S., Clements,J.E. and Ghotbi,Z. (1985) Slow, persistent replication of lentiviruses: Role of tissue macrophages and macrophage precursors in bone marrow., *Proc. Natl. Acad. Sci. USA*, **82**, 7086–7090.

Gendelman,H.E., Narayan,O., Kennedy-Stoskopf,S., Kennedy,P.G.E., Ghotbi,Z., Clements,J.E., Stanley,J., Pezeshkapour,G. (1986) Tropism of sheep lentiviruses for monocytes: Susceptibility to infection and virus gene expression increase during maturation of monocytes to macrophages. *J. Virol.*, **58**, 67–74.

Georgsson,G. and Palsson,P.A. (1971) The histopathology of maedi, a slow viral pneumonia of sheep. *Vet Pathol.*, **8**, 63–80.

Gonda,M.A., Wong-Staal,F., Gallo,R.C., Clements,J.E., Narayan,O. and Gilden,R.V. (1985) Sequence homology and morphologic similarity of HTLV-III and visna virus, a pathogenic lentivirus. *Science*, **227**, 173–177.

Gonda,M.A., Braun,M.J., Clements,J.E., Pyper,J.M. Casey,J.W., Wong-Staal,F., Gallo,R.C., Gilden,R.V. (1986) HTLV-III shares sequence homology with a family of pathogenic lentiviruses. *Proc. Natl. Acad. Sci. USA*, in press.

Gudnadottir,M. (1974) Visna–maedi in sheep. *Prog. Med. Virol.*, **18**, 336–349.

Gudnadottir,M. and Palsson,P. (1967) Transmission of maedi by inoculation of a virus grown in tissue culture from maedi-affected lungs. *J. Infect. Dis.*, **117**, 1–6.

Haase,A.T. (1975) The slow infection caused by visna virus. *Top. Microbiol. Immunol.*, **72**, 101–1565.

Haase,A.T. and Varmus,H.E. (1973) Demonstration of DNA provirus in the lytic growth of visna virus. *Nature New Biol.*, **245**, 237–239.

Haase,A.T., Stowring,L., Narayan,O., Griffin,D. and Price,D.L. (1977) The slow and persistent infection caused by visna virus. The role of host restriction. *Science,* **195**, 175–177.

Haase,A.T., Stowring,L., Harris,J.D., Traynor,B., Ventura,P., Peluso,R. and Brahic,M. (1982) Visna DNA synthesis and the tempo of infections in vitro. *Virology*, **119**, 399–410.

Hess,J.L., Clements,J.E. and Narayan,O. (1985) Cis- and trans-acting transcriptional regulation of visna virus. *Science,* **229**, 482–485.

Johnson,G.C., Barbet,A.F., Klevjer-Anderson,P. and McGuire,T.C. (1983) Preferential immune response to virion surface glycoproteins by caprine arthritis–encephalitis virus-infected goats. *Infect. Immun.* **41**, 657–665.

Kennedy,P.G.E., Narayan,O., Ghotbi,Z., Hopkins,J., Gendelman,H.E. and Clements,J.E. (1985) Persistent expression of Ia antigen and viral genome in visna–maedi virus-induced inflammatory cells: Possible role of lentivirus-induced interferon. *J. Exp. Med.*, **162**, 1970–1982.

Kennedy-Stoskopf,S. and Narayan,O. (1986) Neutralising antibody to visna lentivirus: mechanism of action and possible role in virus persistence. *J. Virol.*, in press.

Klevjer-Anderson,P. and McGuire,T.C. (1982) Neutralizing antibody response of rabbits and goats to caprine arthritis–encephalitis virus. *Infect. Immun.*, **38**, 455–461.

Lin,F.H. and Thormar,H. (1970) Ribonucleic acid dependent deoxyribonucleic acid polymerase in visna virus. *J. Virol.*, **6**, 702–704.

Lutley,R., Petursson,G., Palsson,P.A., Georgsson,G., Klein,J. and Nathanson,N. (1983) Antigenic drift in visna: virus variation during long term infection of kelandic sheep. *J. Gen. Virol.*, **64**, 1433–1440.

Narayan,O. and Cork,L.C. (1985) Lentiviral diseases of sheep and goats: Chronic pneumonia leukoencephalomyelitis and arthritis. *Rev. Infect. Dis.*, **71**, 89–98.

Narayan,O., Griffin,D.E. and Chase,J. (1977a) Antigenic drift of visna virus in persistently infected sheep. *Science,* **197**, 376–378.

Narayan,O., Griffin,D.E. and Silverstein,A. (1977b) Slow virus infection: Replication and mechanisms of persistence of visna virus in sheep. *J. Infect. Dis.*, **135**, 800–806.

Narayan,O., Griffin,D.E. and Clements,J.E. (1978) Virus mutation during 'slow infection'. Temporal development and characterization of mutants of visna virus recovered from sheep. *J. Gen. Virol.*, **41**, 343–352.

Narayan,O., Clements,J.E., Strandberg,J.D., Cork,L.C. and Griffin,D.E. (1980) Biologic characterization of the virus causing leukoencephalitis and arthritis in goats. *J. Gen. Virol.*, **41**, 343–352.

Narayan,O., Clements,J.E., Griffin,D.E. and Wolinsky,J.S. (1981) Neutralizing antibody spectrum determines the antigenic profiles of emerging mutants of visna virus. *Infect. Immun.*, **32**, 1045–1050.

Narayan,O., Wolinsky,J.A., Clements,J.E., Strandberg,J.D., Griffin,D.E. and Cork,L.C. (1982) Slow virus replication: The role of macrophages in the persistence and expression of visna viruses of sheep and goats. *J. Gen. Virol.*, **59**, 345–356.

Narayan,O., Kennedy-Stoskopf,S., Sheffer,D., Griffin,D.E. and Clements,J.E. (1983) Activation of caprine arthritis–encephalitis virus expression during maturation of monocytes to macrophages. *Infect. Immun.*, **41**, 67–73.

Narayan,O., Sheffer,D., Griffin,D.E., Clements,J.E. and Hess,J. (1984) Lack of neutralizing antibodies to caprine arthritis–encephalitis lentivirus in persistently infected goats can be overcome by immunization with inactivated mycobacterium tuberculosis. *J. Virol.*, **49**, 349–355.

Narayan,O., Sheffer,D., Clements,J.E. and Tennekoon,G. (1985) Restricted replication of lentiviruses: Visna viruses induce a unique interferon during interaction between lymphocytes and infected macrophages. *J. Exp. Med.*, **162**, 1954–1969.

Nathanson,N., Panitch,H., Palsson,P.A., Petursson,G. and Georgsson,G. (1976) Pathogenesis of visna. II. Effect of immunosuppression upon the central nervous system lesions. *Lab. Invest.* **35**, 444–451.

Petursson,G., Nathanson,N., Georgsson,G., Panitch,H. and Palsson,P. (1976) Pathogenesis of visna. I. Sequential virologic, serologic, and pathologic studies. *Lab. Invest.*, **35**, 402–412.

Popovic,M., Sarngadharan,M.G., Read,E. and Gallo,R.C. (1984) Detection, isolation, and continuous production of cytopathic retroviruses (HTLV-III) from patients with AIDS and pre-AIDS. *Science,* **224**, 497–500.

Pyper,J.M., Clements,J.E., Molineaux,S.M. and Narayan,O. (1984) Genetic variation among ovine–caprine lentiviruses: Homology between visna virus and caprine arthritis encephalitis virus is confined to the 5′ gag-pol region and a small portion of the env gene. *J. Virol.*, **51**, 713–721.

Roberson,S.M. and Cheevers,W.P. (1984) A physical map of caprine arthritis–encephalitis provirus. *Virology*, **134**, 489–492.

Scott,J.V., Stowring,L., Brahic,M., Haase,A.T., Narayan,O. and Vigne,R. (1979) Antigenic variation in visna virus. *Cell,* **18**, 321–327.

Sigurdsson,B. (1954) Maedi, a slow progressive pneumonia of sheep: an epizootiological and pathological study. *Br. Vet. J.*, **110**, 255–270.

Sigurdsson,B. and Palsson,P.A. (1958) Visna of sheep, a slow demyelinating infection. *Br. J. Exp. Pathol.*, **39**, 519–528.

Sigurdsson,B., Palsson,P.A. and Grissom,H. (1957) Visna, a demyelinating transmissible disease of sheep. *J. Neuropathol. Exp. Neurol.*, **16**, 389–403.

Sigurdsson,B., Thormar,H. and Palsson,P.A. (1960) Cultivation of visna virus in tissue culture. *Arch. ges. Virusforsch.*, **10**, 368–381.

Sonigo,P., Alizon,M., Staskus,K., Klatzman,D., Cole,S., Danos,O., Retzel,E., Tiollais,P., Haase,A.T. and Wain-Hobson,S. (1985) Nucleotide sequence of visna lentivirus: Relationship to the AIDS virus. *Cell*, **42**, 369–382.

Thormar,H., Barshatzky,M.R. and Kozlowski,P.B. (1983) The emergence of antigenic variation is a rare event in long term visna virus infection in vivo. *J. Gen. Virol.*, **64**, 1427–1432.

Wain-Hobson,S., Senigo,P., Donos,O., Cole,S. and Alizon,M. (1985) Nucleotide sequence of the AIDS virus, LAV. *Cell*, **40**, 7–17.

Yaniv,A., Dahlberg,J.E., Tronick,S.R., Chiu,I. and Aaronson,S.A.S. (1985) Molecular cloning of integrated caprine arthritis–encephalitis virus. *Virology*, **145**, 340–345.

CHAPTER 4

Antigenic variation during persistent infections by equine infectious anaemia virus

RONALD C. MONTELARO[1], CHARLES J. ISSEL[2], SUSAN PAYNE[1] AND OLIVIA SALINOVICH[1]

[1]*Department of Biochemistry, and* [2]*Departments of Veterinary Science and Veterinary Microbiology, Louisiana State University and Louisiana Agricultural Experiment Station, Baton Rouge, LA 70803, USA*

Introduction

During the past decade, RNA virus variation and its biological relevance have become increasingly evident, for example in vesicular stomatitis virus, poliovirus and influenza virus (reviewed in Holland *et al.*, 1982). Among retroviruses, where provirus establishment has been widely regarded as *the* mechanism of persistence, a strong case is being developed for the role of genetic and antigenic variations in chronic infections. The occurrence of retrovirus antigenic variation was first detailed for visna virus infections of sheep (Petursson *et al.*, 1976; Narayan *et al.*, 1977; Scott *et al.*, 1979; Griffen *et al.*, 1978; Clements *et al.*, 1980) and this appeared to be an unusual, if not unique, situation. However, subsequent studies of other retrovirus infections, including equine infectious anaemia and acquired immunodeficiency syndrome (AIDS), suggest that variation may be a fundamental property of certain persistent retrovirus infections. The purpose of this presentation is to summarize recent studies on the role and nature of antigenic variation during persistent infections by equine infectious anaemia virus (EIAV).

The Disease

Equine infectious anaemia (EIA) occurs worldwide and is characterized by recurrent eposides of fever, haemolytic anaemia, bone marrow depression, lymphoproliferation, immune-complex glomerulonephritis and persistent viraemia. Although EIA was one of the first diseases to be assigned a viral ('filterable') aetiology (Vallée and Carré, 1904), only within the last several years has progress been made toward understanding the disease at the cellular and molecular level. Clinical, diagnostic, epidemiologic and pathologic aspects of the disease have been reviewed in detail (Henson and McGuire, 1974; Ishii and Ishitani, 1975). Current knowledge relative to virus–host interactions at the molecular level has also been summarized (Issel and Coggins, 1979; Crawford *et al.*, 1978).

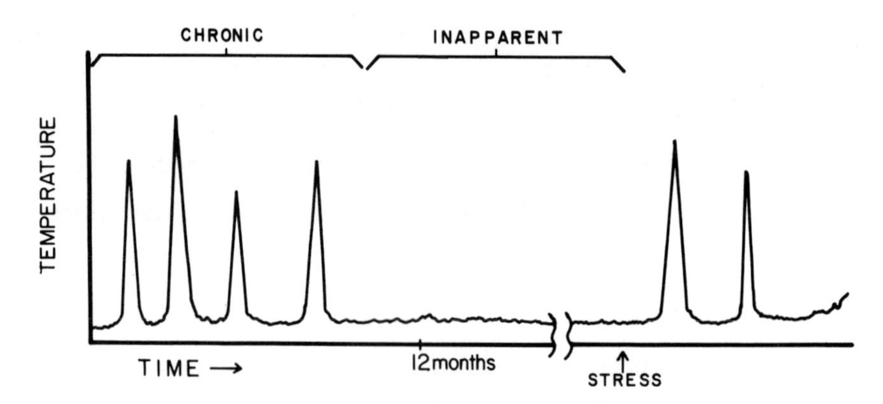

Fig. 1. Schematic representation of the clinical response of a horse after infection by EIAV. Clinical episodes of EIA are represented by peaks of fever above normal animal temperatures. The chronic form of disease is characterized by recurrent clinical episodes which occur at unpredictable intervals during the first year after infection. At this time animals usually no longer display any obvious clinical signs, although recrudescence of disease can be induced years later by stress or the administration of immunosuppressants. The clinical response of animals is variable and depends, in part, on host resistance factors, virus virulence factors and environmental factors (e.g. weather, work, hauling, etc.).

A schematic representation of the clinical response of horses following EIAV infection is presented in *Figure 1*. Acute EIA (fever and haemorrhages) is typically associated with the first exposure to virus and appears to correlate with massive virus replication in, and destruction of, macrophages. Horses in the initial stage of acute EIA are seronegative, but become seropositive as the immune system has time to respond, usually 16–42 days post-infection. The more classical signs of EIAV (weight loss, anaemia and oedema) are seen later during recurring cycles of illness which appear at irregular intervals. This stage of the disease is called chronic EIA. The frequency and severity of clinical episodes decline with time, about 90% occurring within the first year after infection. Horses with chronic EIA, however, usually continue to have unpredictable periodic clinical episodes accompanied by bursts of viraemia. Horses with chronic EIA are seropositive and their T-lymphocytes are responsive to purified, inactivated EIAV (Crawford *et al.*, 1978). The magnitude of the immune response varies during the course of the disease, but the measurement of humoral immune responses to EIAV is complicated by the production of IgG(T) which interferes with various serological tests (Crawford *et al.*, 1978). The majority of naturally infected horses appear to be inapparently infected and no clinical symptoms are observed. However, it has been shown that recrudescence of EIA and detectable viraemia can be induced by certain corticosteroids or by stress, even in some horses which have been free of febrile episodes for years (Kono *et al.*, 1976).

All attempts to vaccinate against EIAV by classic vaccination protocols have proven ineffective (Issel and Coggins, 1979). As a result of these vaccine studies and the development of effective immunodiagnostic tests for exposure to EIAV, it was thought that the spread of this virus could be controlled through local eradication within breeds or within defined geographic regions. Unfortunately, in many areas of the world, the infection rate is too high for these types of control procedures to be effective (Issel and Adams, 1979).

In contrast to the progressive diseases caused by other retroviruses, the periodic nature of EIA is unique and raises a number of intriguing questions concerning the moleuclar biology of virus replication. Moreover, the frequency of the disease episodes in EIA provides a dynamic model for investigating the interaction between host immune systems and viral antigens during a persistent infection, an advantage lacking from most slow virus infections.

The Virus

EIAV is commonly considered a member of the Lentivirus subfamily of Retroviridae, which also includes visna/maedi, progressive pneumonia and zwoegerziekte viruses of sheep, the arthritis–encephalitis virus of goats and the human T-cell lymphotropic (HTLV-III) virus of man (Charman *et al.*, 1976; Stowring *et al.*, 1979; Parekh *et al.*, 1980; Gonda *et al.*, 1985, 1986; Chiu *et al.*, 1985). The relatedness of these viruses, which all cause non-oncogenic chronic diseases, has been established by morphological and serological studies and/or genomic sequence homologies.

The virus particle appears by electron microscopy to be about 100 nm in diameter and contains the usual morphological components characteristic of retroviruses (*Figure 2*) However, in contrast to the icosahedral symmetry observed with the cores of type C and type B oncoviruses, EIAV characteristically displays an oblong core, a property shared by other lentiviruses. Recent studies from our laboratory have provided a detailed analysis of the structural proteins of EIAV including molecular weights, stoichiometry, amino acid compositions, modification patterns (glycosylation and phosphorylation) and isoelectric points (Parekh *et al.*, 1980; Montelaro *et al.*, 1982, 1983). In general, the EIAV proteins are remarkably analogous to those described for murine and avian oncoviruses.

As shown in *Figure 2,* the envelope of the virus contains several hundred copies each

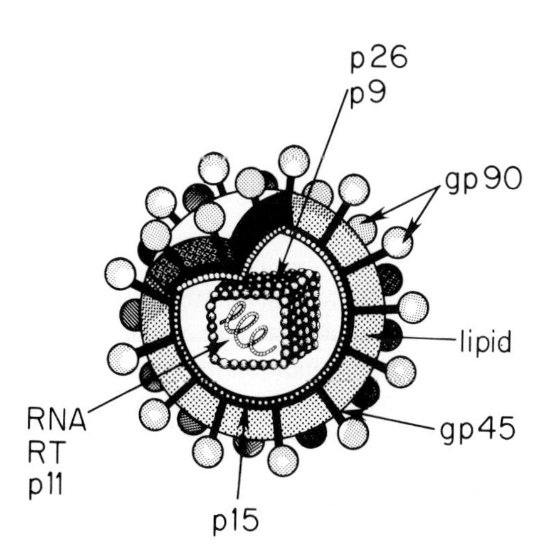

Fig. 2. Structural model for the EIAV virion based on Parekh *et al.* (1980) and Montelaro *et al.* (1982). RT - reverse transcriptase.

of two distinct glycoproteins designated gp90 and gp45. The gp90 component is highly glycosylated and is loosely associated with the virus particle. In contrast, EIAV gp45 is a slightly glycosylated and very hydrophobic protein which apparently spans the lipid bilayer of the virion envelope. The final envelope component is p15, a phosphoprotein which forms a continuous inner coat or mantle layer immediately beneath the lipid bilayer. The envelope glycoproteins gp90 and gp45 are the primary immunogens of the virus during a persistent infection in horses. The antigenicity of the gp90 has been analyzed and the results indicate that the antibody contained in immune serum from persistently infected animals is directed primarily against the peptide portion of gp90, with small, but significant, antibody reactivity specific for the carbohydrate components of the molecule (Montelaro *et al.*, 1984b). In this regard EIAV differs from bovine leukaemia virus gp55 (Portetelle *et al.*, 1980; Schmerr *et al.*, 1981) and Friend murine leukaemia virus gp70 (Bolognesi *et al.*, 1975) where the glycoprotein antigenicity in the natural host has been attributed exclusively to the carbohydrate and peptide moieties of the glycoproteins respectively.

The internal core of EIAV contains 4000–6000 copies each of four major non-glycosylated proteins. EIAV p26, the most abundant structural protein, is the primary component of the core shell and is the antigen employed in commercial diagnostic tests for EIAV antibody in animals (Shane *et al.*, 1984). The other core shell component appears to be the acidic protein, p9. The p11 component is a highly basic protein which binds tightly to the RNA genome of the virus, along with 5–10 copies of reverse transcriptase, forming the ribonucleoprotein complex enclosed by the core shell.

The RNA genome of EIAV is a 60S–70S complex composed of two identical subunits (30S to 40S) (Cheevers *et al.*, 1977). Recently the nucleic acid sequence of EIAV has been partially determined (Chiu *et al.*, 1985; Stephens *et al.*, 1986). The results of these studies indicate a typical genomic retrovirus arrangement of 5′-*gag-pol-env*-3′ with the *gag* sequence coding for the internal virion proteins in the order NH_2-p15-p26-p9-p11-COOH (Stephens *et al.*, 1986). Moreover, comparative sequence studies of EIAV, visna, caprine encephalitis–arthritis and HTLV-III viruses reveal a remarkable degree of homology, suggesting a common progenitor in evolution (Chiu *et al.*, 1985; Stephens *et al.*, 1986). This sequence homology also firmly establishes the distinctness and cohesiveness of the lentivirus subfamily of retroviruses.

The case for variation

Antigenic variation

Antigenic variation, as outlined schematically in *Figure 3*, can be used to explain both the periodic nature of chronic EIA and the failure of classical vaccination procedures. The initial evidence for antigenic variation during persistent infection by EIAV was based on neutralization assays reported by Kono *et al.* (1971, 1973) working in Japan. Although the concept of antigenic variation in EIAV gained wide acceptance among veterinarians, it remained unconfirmed and undefined at a molecular level for over 10 years. This lack of progress undoubtedly resulted from difficulties experienced in establishing reliable animal model systems, in successfully isolating and propagating virus from experimentally infected animals and from the lack of a convenient, reliable neutralization assay for EIAV (Issel and Coggins, 1979). Attempts to distinguish bet-

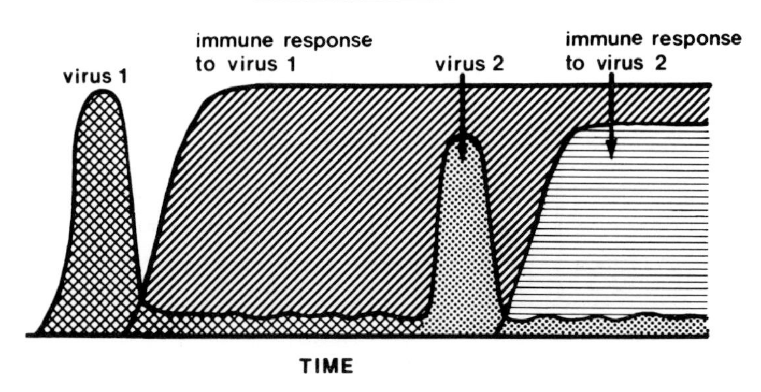

Fig. 3. Model for the recurrent episodes of EIA during the chronic stage of disease. The infecting virus replicates in the horse producing a viraemia and associated clinical signs of EIA. This viraemia is brought under control by immune responses from the infected animal, and clinical manifestations of the disease disappear. However, mutations occurring during the replication of the infecting virus produce novel antigenic stains of virus which are able to circumvent established host immune responses, e.g. neutralizing antibody, resulting in a subsequent viraemia and recurrence of disease. This process can repeat itself.

ween antigenic variants of EIAV by assays such as complement fixation, immunodiffusion and haemagglutination all proved unsuccessful (Sentsui and Kono, 1981). Thus, few laboratories persisted in studying this difficult retrovirus system.

Our research program since 1979 has developed a reliable animal model system in which Shetland ponies inoculated with the cell-adapted Wyoming strain of EIAV reproducibly display characteristic recurrent clinical episodes of EIA (Orrego *et al.*, 1982). Most importantly, EIAV could be successfully recovered and propagated in tissue culture from about 80% of the plasma samples taken from animals during clinical episodes of EIA. These isolates could then be propagated in tissue culture in quantities sufficient for detailed immunological and biochemical comparisons.

Figure 4 summarizes the clinical history of a representative experimentally infected pony involved in the serial passage experiment. The animal experienced four clinical episodes of EIA over a 5-month observation period, with individual episodes spaced by an average of 4–5 weeks. Virus isolates (designated P31-1, P3-2, P3-3 and P3-4, respectively) were successfully obtained from end-point dilutions of plasma taken from the pony during each febrile episode, whereas no virus could be recovered from plasma samples taken between febrile episodes. These results indicate low levels of virus replication during the persistent infection, except for bursts of viraemia associated with disease episodes.

According to the model for EIAV persistence (*Figure 3*), the isolates obtained from the different clinical episodes should be antigenically distinct from each other. This prediction was shown to be correct by a comparison of the neutralization properties of the viurs isolates with serum samples taken from the pony at various time points (Salinovich *et al.*, 1986). The results of these studies indicated that each virus isolate was effectively neutralized by serum samples taken *after* the febrile episode of origin, demonstrating that the pony is able to produce neutralizing antibody in response to EIAV infection. However, none of the virus isolates was neutralized by serum samples taken

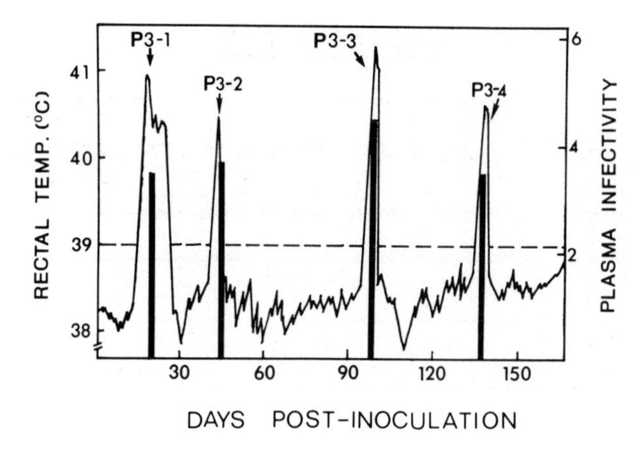

Fig. 4. Clinical history of an experimentally infected pony showing disease episodes from which virus isolates were recovered. All sustained rectal temperature recordings above 39°C are considered abnormal and designated as febrile episodes. Vertical bars indicate plasma viraemia levels at the time of isolate recovery, as determined by titration on foetal equine kidney cells, and expressed as $\log_{10}$ $TCID_{50}$ per 0.5 ml (Salinovich *et al.*, 1986).

prior to the febrile episode of origin, indicating that in each case the existing antibody is ineffective in preventing the emergence of the viraemia and associated pathology. For example, serum taken 18 days after the first febrile episode (*Figure 4*) neutralized the P3-1 isolate, but failed to neutralize the later three isolates. In contrast, serum taken 3 weeks after the third febrile episode neutralized the P3-1, P3-2 and P3-3 isolates; only P3-4 was not affected by the serum. Thus, these observations demonstrate that distinct antigenic variants of EIAV are associated with each disease episode in the sequential series. Moreover, the data distinguishes EIAV from visna where 1–2 years are required for the emergence of antigenic variants (Clements *et al.*, 1980; Scott *et al.*, 1979) and from persistent infections by caprine arthritis–encephalitis virus (CAEV) in which significant levels of neutralizing antibody are not detectable (Narayan *et al.*, 1982).

The antigenic relatedness of the four isolates of EIAV has also been compared by Western blot analyses utilizing both polyclonal immune serum from infected animals and specific monoclonal antibodies produced against the gp90 component of the prototype strain of EIAV (*Figure 5*). Western blots with a high titre reference equine immune serum revealed that the primary immunogens during persistent infection by EIAV are the surface glycoproteins gp90 and gp45 (*Figure 5A*). A minor antibody titre is also detectable against the major core protein p26, presumably resulting from disruption of the virus. The data in *Figure 5A* also demonstrate that the four isolates of EIAV and the prototype strain display extensive gp90 and gp45 cross-reactivities, although these viruses are distinct in neutralization assays. This pattern of reactivity suggests that immune serum from a persistently infected animal contains large quantities of antibody that will bind to a wide spectrum of virus isolates, but very specific populations of neutralizing antibody. A previous report indicating the presence of infectious virus–antibody complexes in the blood of infected animals (McGuire *et al.*, 1972) apparently correlates with the Western blot results presented here.

In contrast to the broad reactivity of the equine immune serum, monoclonal antibody

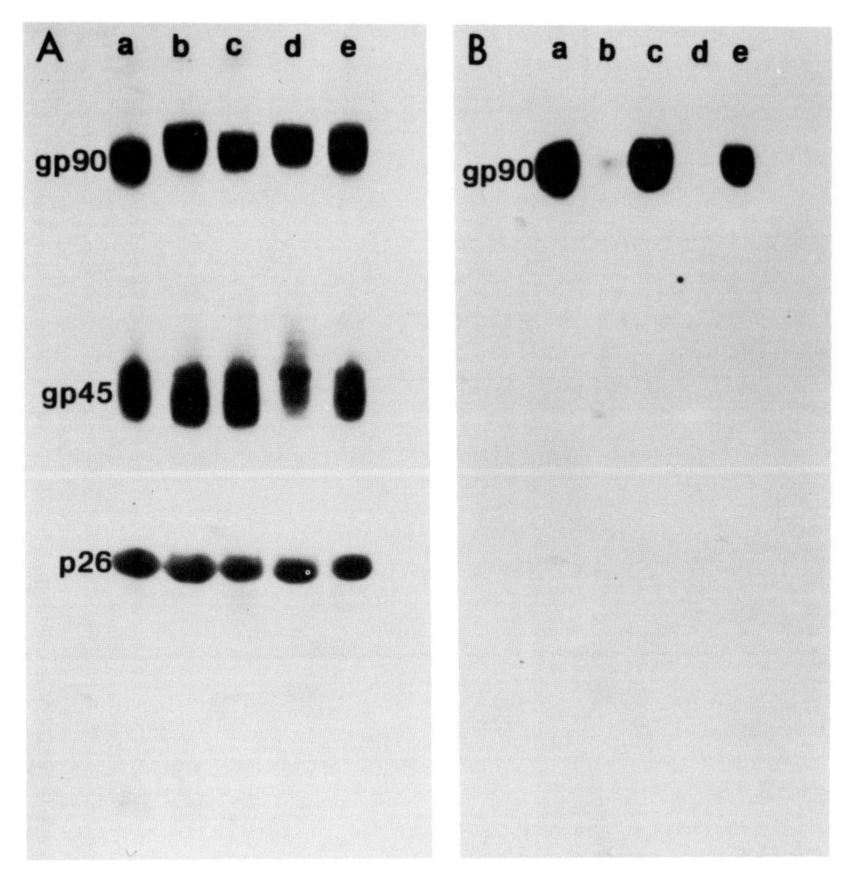

Fig. 5. Immunoblot analysis of purified test strains of EIAV: prototype (**a**), P3-1 (**b**), P3-2 (**c**), P3-3 (**d**) and P3-4 (**e**). **Panel A**, 70 μg samples of each virus isolate were first analyzed by SDS–PAGE, transferred to nitrocellulose, and treated with reference horse immune serum to identify the immunogenic virion proteins. **Panel B**, 50 μg samples of each virus isolate were analyzed by SDS–PAGE and transferred as in **panel A**, then treated with a monoclonal antibody generated against gp90 from prototype EIAV.

preparations are able to distinguish between isolates of EIAV in Western blot analysis (Salinovich *et al.*, 1986). The profile in *Figure 5B* is representative in that the monoclonal antibody reacts with prototype, P3-2 and P3-4 gp90 molecules, but fails to react with gp90 of the P3-1 and P3-3 isolates. In surveying a panel of monoclonal antibodies it has been observed that both neutralizing and non-neutralizing monoclonal antibody preparations will differentiate between EIAV isolates. In contrast, monoclonal antibody preparations which react with *all* EIAV isolates tested thus far are always non-neutralizing. The patterns of monoclonal antibody reactivity observed in Western blots are also observed in radioimmunoassay procedures using [^{125}I]gp90 from different virus isolates. These results demonstrate the variation of single epitopes during persistent infection by EIAV and suggest that both neutralizing and non-neutralizing monoclonal antibody preparations are able to distinguish between variants of EIAV. In contrast, only non-neutralizing antibodies appear to identify epitopes conserved among isolates of EIAV.

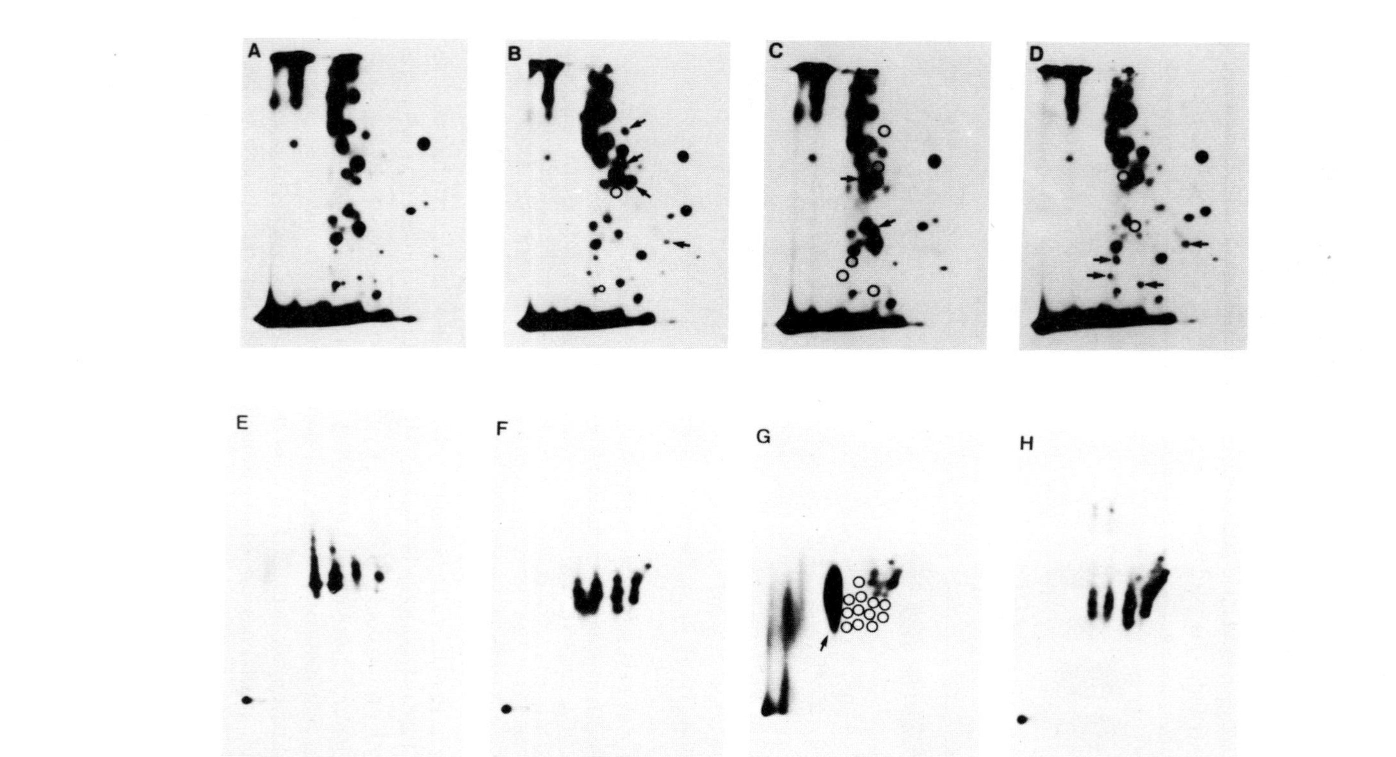

Fig. 6. Tryptic peptide and glycopeptide maps of [^{125}I]gp90 isolated from EIAV strain P3-1 (**A** and **E**), P3-2 (**B** and **F**), P3-3 (**C** and **G**) and P3-4 (**D** and **H**). The two-dimensional analysis consisted of electrophoresis in the horizontal direction and chromatography in the vertical direction. Samples were applied in the lower left corner. **Panels A – D** are maps of non-glycosylated peptides of gp90 from each isolate. Arrows (additions) and circles (deletions) indicate peptide differences in each map when compared with that isolate's immediate predecessor. **Panels E – H** are maps of glycosylated peptides isolated from gp90 from each virus stain. In view of probable microheterogeneity of the carbohydrate moieties, vertical families of glycopeptides which possess the same electrophoretic mobility were utilized in comparisons instead of individual glycopeptides. Arrows (additions) and circles (deletions) in **panel G** indicate major differences in this map of P3-3 gp90 glycopeptides as compared with maps from the other three isolates.

Protein structural variation

A surprising finding from the Western blot profiles (*Figure 5*) is that the glycoprotein components of the different EIAV isolates display slight, but highly reproducible, variations in their respective electrophoretic mobilities in sodium dodecylsulphate–polyacrylamide gel electrophoresis (SDS–PAGE). The internal virion proteins of the different isolates, e.g. p26, always display identical electrophoretic mobilities. Hence these specific alterations in glycoprotein electrophoretic mobility suggest structural variations in the respective gp90 and gp45 components of the EIAV isolates, perhaps reflecting the antigenic variations observed in neutralization assays and Western blot analyses.

The structural variation suggested by the alterations in electrophoretic mobility can be demonstrated by comparative two-dimensional peptide and glycopeptide analysis of the glycoprotein components of the EIAV isolates (Montelaro *et al.*, 1984a; Salinovich *et al.*, 1986). *Figure 6* presents a comparison of the gp90 molecules P3-1, P3-2, P3-3 and P3-4 isolates of EIAV, while *Figure 7* presents an analogous comparison of the respective gp45 components of these isolates. To obtain a systematic comparison, peptide additions and deletions are indicated based on a comparison of each isolate with its immediate predecessor. In this manner, P3-2 is compared with P3-1, P3-3 with P3-2 and P3-4 with P3-3. The results of each comparison of peptide maps reveal differences (peptide additions and deletions) in each pair of gp90 (*Figure 6A–D*) and gp45 (*Figure 7A–D*) molecules examined. The maps obtained for the *glycosylated* tryptic peptides of the gp90 (*Figure 6E–H*) and of the gp45 (*Figure 7E–H*) reveal two distinct patterns of glycosylation in each glycoprotein among the isolates examined. Stains P3-1, P3-2 and P3-4 display identical glycosylation patterns for gp90, while P3-3 exhibits a strikingly different glycosylation pattern. Similarly, the P3-1 and P3-3 gp45 molecules display an identical glycosylation pattern, which contrasts markedly with the pattern displayed by both the P3-2 and P3-4 gp45 molecules.

Taken together, the results from these peptide and glycopeptide mapping studies demonstrate that the glycoproteins of these isolates of EIAV contain structurally different gp90 and gp45 components. Each isolate contains a unique amino acid sequence, as demonstrated by unique peptide maps, and certain isolates also differ in their patterns of glycosylation, as indicated by the variations in glycopeptide maps. These structural variations may correlate with the observed antigenic variation.

In contrast to the variations found in the glycoproteins of the isolates of EIAV, peptide mapping of the internal structural proteins indicates no variations in these proteins among the isolates examined (Salinovich *et al.*, 1986). Representative peptide maps of the p26 and p15 components of the P3-1, P3-2, P3-3 and P3-4 stains of EIAV are presented in *Figure 8*. The identical maps obtained for p26 and p15 molecules of each isolate indicate that the sequence of these proteins is not altered during persistent infection. A similar conservation of structure is indicated by the peptide maps of the other internal proteins (data not shown). Thus the structural variation observed in EIAV during persistent infection is evidently specific for the virion glycoproteins, which are therefore the target for immune selection mechanisms.

Genomic variation

Having established the occurrence of antigenic variations and the specific alteration of glycoproteins of EIAV during persistent infections, the RNA genomes of the virus

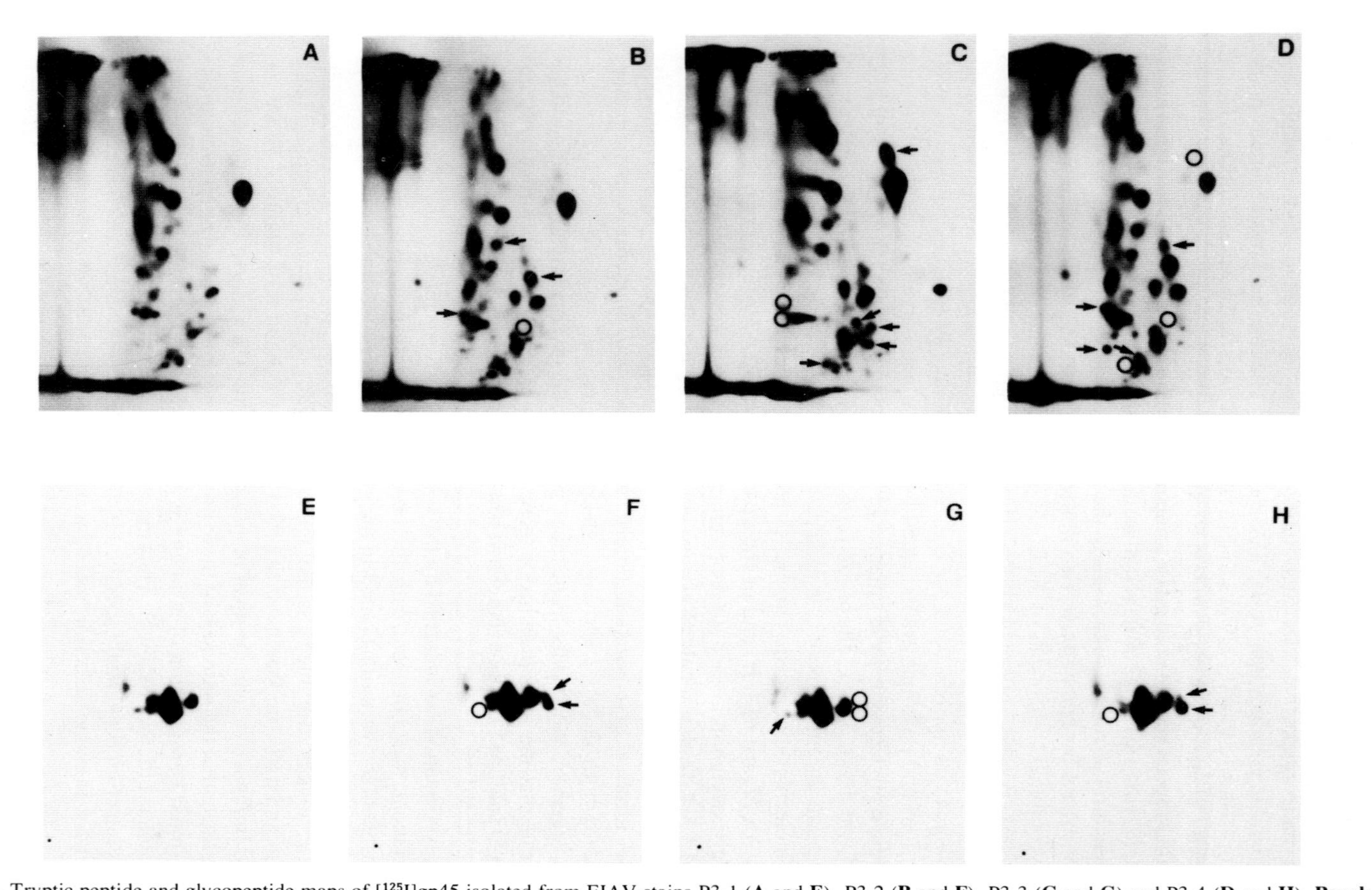

Fig. 7. Tryptic peptide and glycopeptide maps of [^{125}I]gp45 isolated from EIAV stains P3-1 (**A** and **E**), P3-2 (**B** and **F**), P3-3 (**C** and **G**) and P3-4 (**D** and **H**). **Panels A−D**, peptide analysis and comparisons are as described in *Figure 6*. **Panels E−H**, arrows (additions) and circles (deletions) represent glycopeptide differences in each map when compared with that isolate's immediate predecessor, as indicaed for the gp45 peptide maps.

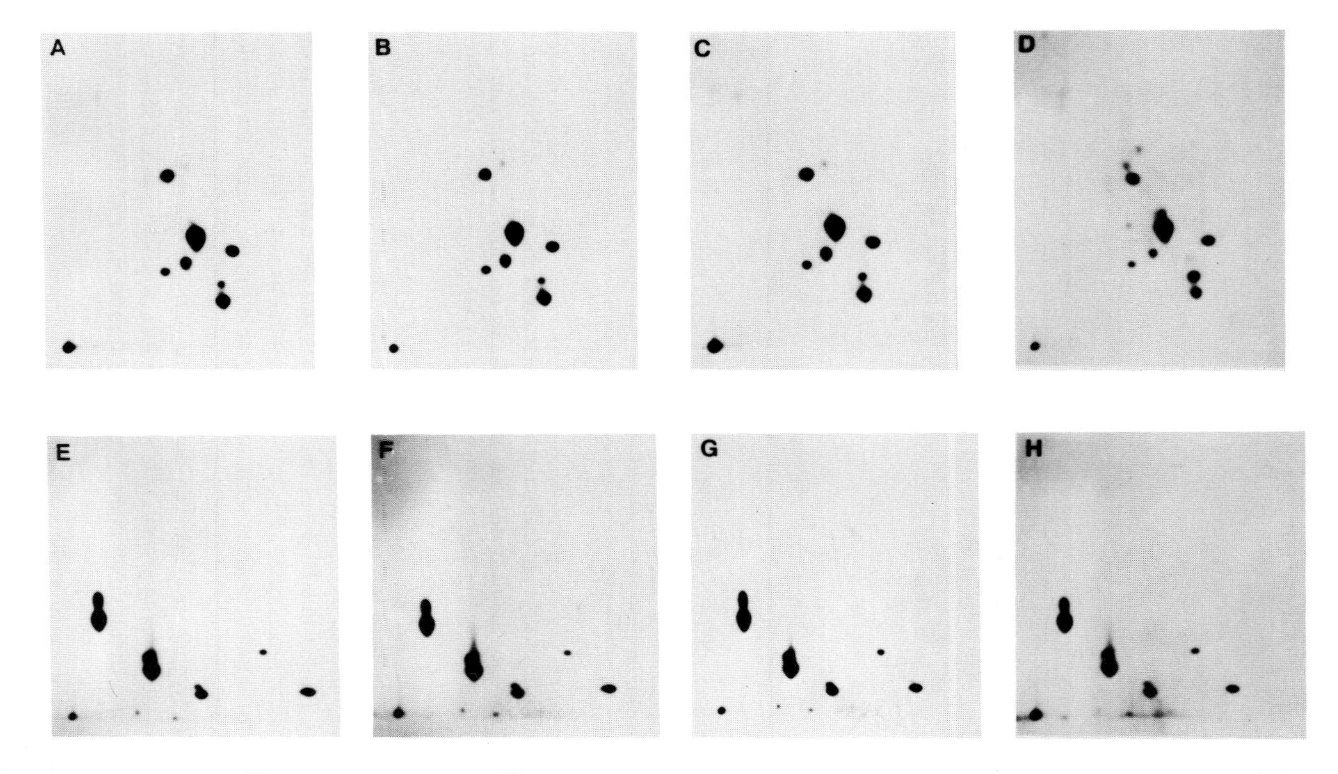

Fig. 8. Tryptic peptide maps of [^{125}I]p26 (**panels A–D**) and [^{125}I]p15 (**panels E–G**) isolated from EIAV strains P3-1, P3-2, P3-3 an P3-4, respectively. Peptide analysis and comparisons are as described in *Figure 5*.

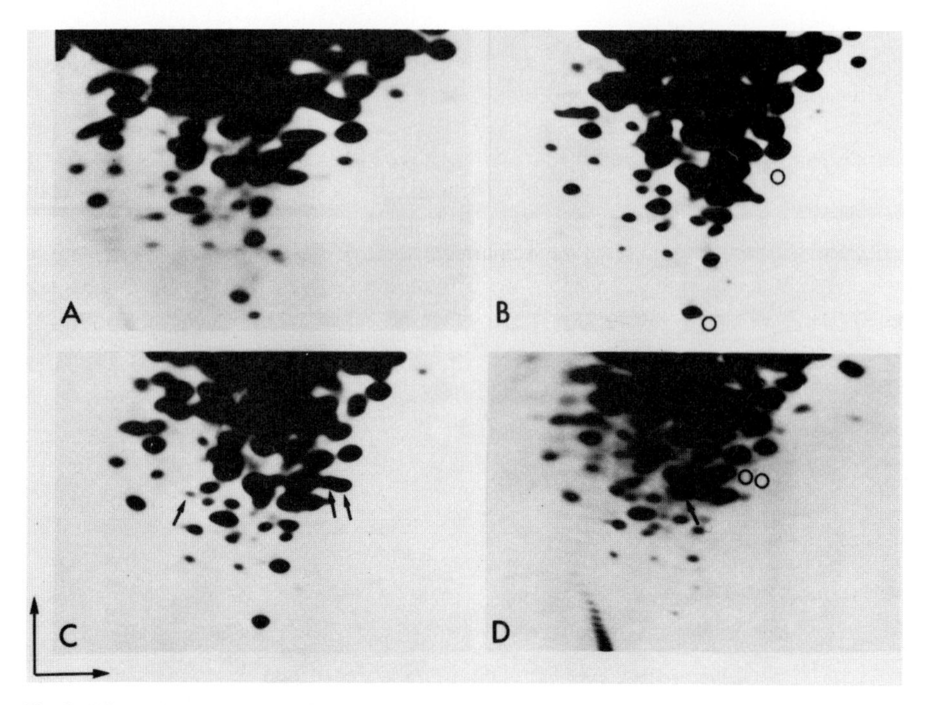

Fig. 9. Oligonucleotide maps of EIAV 70S genomic RNA from third passage isolates P3-1 (**a**), P3-2 (**b**), P3-3 (**c**) and P3-4 (**d**). The direction of electrophoresis in the first dimension (8% polyacrylamide, pH 3.25) was from left to right and in the second dimension (23% polyacrylamide, pH 8.2) from bottom to top. Arrows (additions) and open circles (deletions) indicate diferences in each pattern when compared with its immediate predecessor, as was done in the peptide maps.

isolates were compared by oligonucleotide mapping procedures to determine whether the changes in glycoprotein structure could be detected at the level of RNA and to see whether these changes result from point mutations or recombinational events. *Figure 9* presents representative oligonucleotide maps for EIAV isolates P3-1, P3-2, P3-3 and P3-4. Comparisons of the maps reveal highly similar, but unique, sets of oligonucleotides for each of the four isolates examined. The map of EIAV P3-2 compared with that of its immediate predecessor, P3-1, shows two oligonucleotide deletions (*Figure 9B*). The oligonucleotide map for isolate P3-3 has three additional oligonucleotides when compared with P3-2, while P3-4 has two deletions and one addition when compared with its immediate predecessor, P3-3. It is interesting to note that while a relatively small number of oligonucleotides seem to be variable, no two maps are identical, as observed with the corresponding glycoprotein peptide maps. Moreover, the variant oligonucleotides typically map in the envelope region of the virus genome when localized by oligonucleotide mapping procedures (Salinovich *et al.*, 1986); only a small number of variant oligonucleotides map in the *gag* (internal protein) or *pol* (reverse transcriptase) regions of the genome.

Genomic variation can be attributed to either point mutations or recombinational events during virus replication. However, the relatively low level of genomic variation and the lack of endogenous EIAV genes in host cells (Rice *et al.*, 1978) favours the model in which EIAV undergoes genomic point mutations during replication, presumably resulting from unfaithful reverse transcription which must occur in the absence of editing and repair enzymes (Salinovich *et al.*, 1986; Payne *et al.*, 1984). It is interesting to note that the total number of oligonucleotide alterations observed between the first virus isolate (P3-1) and the last virus isolate (P3-4) is no greater than the total number of alterations observed between any two consecutive isolates. There does not appear to be an accumulation of point mutations. Instead, the observed changes are evidently due to the presence or absence of a specific, limited subset of oligonucleotides, while the majority of the oligonucleotides remain constant. Unique oligonucleotide patterns then appear to be obtained by various permissible combinations of this oligonucleotide subset. Although point mutations probably generate the original variations, recombination between subsequent variants could further propagate variation during a persistent infection.

Thus the oligonucleotide mapping data indicate that EIAV undergoes constant genomic variations during replication in persistently infected animals. Some of these genomic variations result in altered glycoprotein structures, some of which, as antigenic variants, circumvent established host immune responses, resulting in a burst of viraemia and recrudescence of disease.

Perspectives on EIAV variation

The studies decribed here with EIAV indicate that the virus is able to undergo relatively rapid genomic and antigenic variations during a persistent infection process. This property of variation evidently explains the unique periodic nature of EIA and the lack of success in using classic vaccination procedures to protect against virus infection. However, the studies also raise a number of important questions about the molecular biology of this virus, including the extent of variation possible and the number of biochemical properties of variant and conserved epitopes on the virion glycoproteins.

How many serotypes of EIAV exist in nature? Are there three serotypes as for poliovirus or over 100 serotypes as for rhinoviruses? The answer is not yet known, and will depend partially on the cirteria used for comparison, e.g. neutralization properties, reactivity in immunoassays such as Western blots, radioimmunoassay, etc. Thus far in our survey of isolates generated in a serial passage, as many as six neutralization-distinct isolates have been recovered from a single animal. In addition, in comparing over 20 isolates by peptide and oligonucleotide mapping procedures, we have failed to identify two identical isolates. Thus it seems safe to assume that a relatively large number of distinct serotypes of EIAV exist in nature.

What is the extent of genomic and amino acid sequence variation between isolates? Precise information must await genomic sequencing analysis of a variety of isolates, and these experiments are in progress. It should be emphasized here, however, that the peptide and oligonucleotide mapping procedures employed in the current studies provide only a minimum estimate of total structural alterations. In the case of peptide mapping, only those peptides containing tyrosyl and/or histidyl residues susceptible to iodination are monitored; peptides lacking these residues, and perhaps undergoing variation, do not appear in the maps. Similarly, the 45–50 unique high molecular weight oligonucleotides examined for comparison of different virus strains probably comprise only 10–15% of the total EIAV genome (Clements *et al.*, 1980; Pederson and Haseltine, 1980). Thus, the actual level of glycoprotein and genomic variation between isolates is certainly more extensive than revealed by the analytical procedures employed here.

Is the variation observed in EIAV atypical for a retrovirus? The fact that variation is observed between virus isolates, recovered only 4–8 weeks apart from the same animal, indicates that EIAV is a highly mutable virus. However this property is not limited to EIAV. Indeed, antigenic and genomic variations during persistent retrovirus infections were first reported for visna virus, but their importance in virus persistence has been questioned following recent serological analyses. These reported either the presence of parental strains of visna virus long after new variants had appeared in experimentally infected sheep (Lutley *et al.*, 1983) or the lack of antigenic variants in some sheep even during advanced stages of the disease (Thormar *et al.*, 1983). More recently, isolates of HTLV-III (AIDS virus) have been shown to exhibit nucleic acid sequence variation, but only in the envelope gene region (Hahn *et al.*, 1984; Ratner *et al.* 1985; Sanchez-Pescador *et al.*, 1985; Wong-Staal *et al.*, 1985). EIAV, visna virus and HTLV-III have all been assigned to the lentivirus subfamily of retroviruses (Chiu *et al.*, 1985; Stephens *et al.*, 1986; Gonda, *et al.*, 1986), strengthening the case that the capacity for antigenic variation is a common characteristic of this subfamily. Moreover, the increased number of reports of frequent spontaneous genetic variation in a variety of avian and murine oncoviruses (Kawai and Hanafusa, 1973; Zarlin and Temin, 1976; Lee *et al.*, 1981; O'Rear and Temin, 1982; Darlix and Spahr, 1983; Shtivelman *et al.*, 1984) may indicate a characteristic property of retroviruses whose biological relevance has been overlooked. There is little information on the possible role of the variation in tumour development in animals, although glycoprotein antigenic variation has recently been reported for bovine leukaemia virus (Bruck *et al.*, 1984).

Concluding remarks

Considering all the available retrovirus systems, the dynamic nature of EIAV infections and virus variation clearly offers a unique model for studying the mechanisms and nature of antigenic variation in a persistent retrovirus infection where the process is of biological importance. Studies are currently underway to identify and characterize the epitopes of the EIAV glycoproteins and to determine the range and chemical nature of their variation. This information should, hopefully, serve as a foundation for addressing the fundamental question of how one protects against a virus that is able to change under immune pressures.

Although the development of an effective vaccine against EIAV appears to be a formidable if not impossible task, new technologies such as synthetic peptide vaccines may provide an approach if a conserved region(s) of the virion can be identified. A possible candidate for a conserved site to be used as an immune target is the receptor binding portion of the EIAV gp90 molecule. The challenge is there, not only for EIAV, but for other persistent virus infections of both human and veterinary concern.

Acknowledgements

The authors express their appreciation to J. Casey and N. Rice for providing unpublished data from their laboratories. This research is supported by funds from the Louisiana Agricultural Experiment Station, The Louisiana State University School of Veterinary Medicine, National Institutes of Health Grant CA-38851, and United States Department of Agriculture Grant 85-CRCR-1-1804.

References

Bolognesi,D., Collins,J., Leis,J., Moennig,V., Schafer,W. and Atkinson,P. (1975) Role of carbohydrate in determining the immunochemical properties of the major glycoprotein (gp71) of Friend murine leukemia virus. *J. Virol.*, **16**, 1453–1463.

Bruck,C., Resonnet,N., Portetelle,D., Cleuter,Y., Mammerickx,M., Burny,A., Mamoun,R., Guilleman,B., Van der Maaten,M. and Ghysdael,J. (1984) Biologically active epitopes of bovine leukemia virus glycoprotein gp51: their dependence on protein glycosylation and genetic variability. *Virology*, **136**, 20–31.

Charman,H.P., Bladen,S., Gilden,R.V. and Coggins,L. (1976) Equine infectious anemia virus: evidence favoring classification as a retrovirus. *J. Virol.*, **19**, 1073–1079.

Cheevers,W., Archer,B. and Crawford,T. (1977) Characterization of RNA from EIAV. *J. Virol.*, **24**, 482–497.

Chiu,I., Yaniu,A., Dahlberg,J. Gazit,A., Skuntz,S., Tronick,S. and Aaronson,S. (1985) Nucleotide sequence evidence for relationship of AIDS retrovirus to lentiviruses. *Nature*, **317**, 366–368.

Clements,J.E., Pederson,F.S., Narayan,O. and Haseltine,W.A. (1980) Genomic changes associated with antigenic variation of visna virus during persistent infection. *Proc. Natl. Acad. Sci. USA*, **77**, 4454–4458.

Crawford,T.B., Cheevers,W.P., Klevjer-Anderson,P. and McGuire,T.C. (1978) Equine infectious anemia: virion characteristics, virus-cell interactions and host responses. In *Persistent Viruses. ICN-UCLA Symposium on Molecular and Cellular Biology, vol. 11.* Stevens,J., Todaro,G. and Fox,C.F. (eds), Academic Press, New York, pp. 727–749.

Darlix,S. and Spahr,P. (1983) High spontaneous mutation rate of Rous sarcoma virus demonstrated by direct sequencing of the RNA genome. *Nucleic Acids Res.*, **11**, 5953–5967.

Gonda,M.A., Wong-Staal,F., Gallo,R.C., Clements, J.E., Narayan,O. and Gilden,R.V. (1985) Sequence homology and morphologic similarity of HTLV-III and visna virus, a pathogenic lentivirus. *Science*, **227**, 173–177.

Griffen,D.E., Narayan,O. and Adams,R.J. (1978) Early immune response in visna, a slow viral disease of sheep. *J. Infect. Dis.*, **138**, 340–350.

Hahn,B.H., Shaw,G.M., Arya,S.K., Popavic,M., Gallo,R.C. and Wong-Staal,F. (1984) Molecular cloning and characterization of the HTLV-III virus associated with AIDS. *Nature*, **312**, 166–169.

Henson,J. and McGuire,T. (1974) Equine infectious anemia. *Prog. Med. Virol.*, **18**, 143–159.

Holland,J., Spindler,K., Horodyski,F., Grabau,E., Nichol,S. and Van de Pol,S. (1982) Rapid evolution of RNA genomes. *Science*, **215**, 1577–1585.

Ishii,S. and Ishitani,R. (1975) Equine infectious anemia. *Ad. Vet. Sci. Comp. Med.*, **19**, 195–198.

Issel,C.J. and Adams,W.V. (1979) Serologic survey for EIAV in Louisiana. *J. Am. Vet. Med. Assoc.*, **174**, 286–288.

Issel,C. and Coggins,L. (1979) Equine infectious anemia: current knowledge. *J. Am. Vet. Med. Assoc.*, **174**, 286–288.

Kawai,S. and Hanafusa,H. (1973) Isolation of a defective mutant of avian sarcoma virus. *Proc. Natl. Acad. Sci. USA*, **70**, 3493–3497.

Kono,Y., Kobayaski,K. and Fukunaga,Y. (1971) Serological comparisons among strains of EIAV. *Arch. Virol.*, **34**, 202–208.

Kono,Y., Kobayashi,K. and Fukunaga,Y. (1973) Antigenic drift of equine infectious anemia virus in chronically infected horses. *Arch. Gesamte Virusforsch.*, **41**, 1–10.

Kono,Y., Hirasawa,K., Fukunaga,Y. and Taniguchi,T. (1976) Recrudescence of EIA by treatment with immunosuppressive drugs. *Natl. Inst. Anim. Health. Q (Tokyo)*, **16**, 8–15.

Lee,W., Nunn,M. and Duesberg,P.H. (1981) *src* genes of ten Rous sarcoma virus strains, including two reportedly transduced from the cell, are completely allelic; putative markers of transduction are not detected. *J. Virol.*, **39**, 758–776.

Lutley,R., Petursson,G., Palisson,P.A., Georgsson,G., Klein,J. and Nathanson,N. (1983) Antigenic drift in visna: virus variation during long-term infection of Icelandic sheep. *J. Gen. Virol.*, **64**, 1433–1440.

McGuire,T., Crawford,T. and Henson,J. (1972) Equine infectious anemia: detection of infectious virus-antibody complexes in the serum. *Immunol. Commun.*, **1**, 545–551.

Montelaro,R.C., Lohrey,N., Parekh,B., Blakeney,E.W. and Issel,C.J. (1982) Isolation and comparative biochemical properties of the major internal polypeptides of equine infectious anemia virus. *J. Virol.*, **42**, 1029–1038.

Montelaro,R.C., West,M. and Issel,C.J. (1983) Isolation of equine infectious anemia virus glycoproteins. Lectin affinity chromatography procedures for high avidity glycoproteins. *J. Virol. Methods*, **6**, 337–346.

Montelaro,R.C., Parekh,B., Orrego,A. and Issel,C.J. (1984a) Antigenic variation during persistent infection by equine infectious anemia virus, a retrovirus. *J. Biol. Chem.*, **259**, 10539–10544.
Montelaro,R.C., West,M. and Issel,C.J. (1984b) Antigenic reactivity of the major glycoprotein of equine infectious anemia virus, a retrovirus. *Virology*, **136**, 368–374.
Narayan,O., Griffen,D.E. and Chase,J. (1977) Antigenic shift in visna virus in persistently infected sheep. *Science*, **197**, 376–378.
Narayan,O., Sheffer,D., Griffin,D., Clements,J. and Hess,J. (1982) Lack of neutralizing antibodies to caprine arthritis-encephalitis lentivirus in persistently infected goats can be overcome by immunization with inactivated mycobacterium tuberculosis. *J. Virol.*, **49**, 349–355.
O'Rear,J. and Temin,H. (1982) Spontaneous changes in nucleotide sequence in proviruses of spleen necrosis virus, an avian retrovirus. *Proc. Natl. Acad. Sci. USA*, **79**, 1230–1234.
Orrego,A., Issel,C.J., Montelaro,R.C. and Adams,W.V. (1982) Virulence and in vitro growth of a cell-adapted strain of equine infectious anemia virus after serial passage in ponies. *Am. J. Vet. Res.*, **43**, 1556–1560.
Parekh,B., Issel,C.J. and Montelaro,R.C. (1980) Equine infectious anemia virus, a putative lentivirus, contains polypeptides analogous to prototype-C oncornaviruses. *Virology*, **107**, 520–525.
Payne,S., Parekh,B., Montelaro,R.C. and Issel,C.J. (1984) Genomic alterations associated with persistent infections by equine infectious anemia virus, a retrovirus. *J. Gen. Virol.*, **65**, 1395–1399.
Pederson,F.S. and Haseltine,W.A. (1980) Analysis of the genome of an endogenous, ecotropic retrovirus of the AKR strain of mice: micromethod for detailed characterization of high-molecular-weight RNA. *J. Virol.*, **33**, 349–365.
Petursson,G., Nathanson,N., Georgsson,G. Panitch,H. and Palsson,P. (1976) Pathogenesis of visna. I. Sequential virologic, serologic, and pathologic studies. *Lab. Invest.*, **35**, 402–412.
Portetelle,D., Bruck,C., Mammerickx,M. and Burny,A. (1980) In animals infected by bovine leukemia virus (BLV) antibodies to envelope glycoprotein gp51 are directed against the carbohydrate moiety. *Virology*, **105**, 223–233.
Ratner,L., Haseltine,W.A., Patarca,R., Livak,K.J., Starcich,B., Josephs,S.F., Doran,E.R., Rafalski,J.A., Whitehorn,E.A., Baumeister,K., Ivanoff,L. Petteway,S.R.,Jr., Pearson,M.L. Lautenberger,J.A., Papas,T.S., Ghrayeb,J., Chang,N.T., Gallo,R.C. and Wong-Staal,F. (1985) Complete nucleotide sequence of the AIDS virus, HTLV-III. *Nature*, **313**, 277–284.
Rice,N.R., Simek,S., Ryder,O.A. and Coggins,L. (1978) Detection of proviral DNA in horse cells infected with equine infectious anemia virus. *J. Virol.*, **26**, 577–583.
Salinovich,O., Payne,S., Montelaro,R., Hussain,K., Issel,C. and Schnorr,K. (1986) The rapid emergence of novel antigenic and genetic variants of EIAV during persistent infection. *J. Virol.*, **57**, 71–80.
Sanchez-Pescador,R., Power,M.D., Barr,P.J., Steimer,K.S., Stempien,M.M., Brown-Shimer,S.L., Gee,W.W., Renard,A., Randolph,A., Levy,J.A., Dina,D. and Luciw,P.A. (1985) Nucleotide sequence and expression of an AIDS-associated retrovirus (ARV-2). *Science*, **227**, 484–492.
Schmerr,M.J.F., Miller,J.M. and Van der Maaten,M.J. (1981) Antigenic reactivity of a soluble glycoprotein associated with bovine leukemia virus. *Virology*, **109**, 431–434.
Scott,J.V., Haase,A.T., Narayan,O. and Vinge,R. (1979) Antigenic variation in visna virus. *Cell*, **18**, 321–327.
Sentsui,H. and Kono,Y.. (1981) Hemagglutination of several strains of equine infectious anemia virus. *Arch. Virol.*, **67**, 75–84.
Shane,B.S., Issel,C.J. and Montelaro,R.C. (1984) Enzyme-linked immunosorbent assay for detection of equine infectious anemia virus p26 antigen and antibody. *J. Clin. Microbiol.*, **19**, 351–355.
Shtivelman,E., Zakut,R. and Canaani,E. (1984) Frequent generation of nonrescuable reorganized Moloney murine sarcoma viral genomes. *Proc. Natl. Acad. Sci. USA*, **81**, 294–298.
Stephens,R., Casey,J. and Rice,N. (1986) Nucleotide sequence of the *gag* and *pol* genes of EIAV: relatedness to visna and to AIDS virus. *Science*, **231**, 589–594.
Stowring,L., Haase,A.T. and Charman,H.P. (1979) Serological definition of the lentivirus group of retroviruses. *J. Virol.*, **29**, 523–528.
Thormar,H., Barshatzky,M., Arnesen,K. and Kozlowski,P. (1983) The emergence of antigenic variants is a rare event in long term visna virus infection *in vivo*. *J. Gen. Virol.*, **64**, 1427–1432.
Vallée,H. and Carré,H. (1904) Sur la nature infectieuse de l'anémie du cheval. *C.R. Acad. Sci.*, **139**, 331–333.
Wong-Staal,F., Shaw,G., Hahn,B., Salahuddin,S. Popovic,M., Markham,P., Redfield,R. and Gallo,R. (1985) Genomic diversity of human T-lymphotropic virus type III (HTLV-III). *Science*, **229**, 759–762.
Zarlin,D. and Temin,H. (1976) High spontaneous mutation rate of avian sarcoma virus. *J. Virol.*, **17**, 74–84.

CHAPTER 5

The genetic basis of phase and antigenic variation in bacteria

J.R. SAUNDERS

Department of Microbiology, University of Liverpool, Liverpool L69 3BX, UK

Introduction

Losses and/or alteration of observable characteristics such as colonial morphology or virulence properties are frequently observed when pathogenic bacteria are cultivated in the laboratory. These phenotypic changes in bacteria grown *in vitro* are paralleled by the dramatic ability of certain pathogens to alter the surface characteristics that they present to their environment(s). The genetic bases of most of the events that have been observed have never been investigated. However, sufficient information is now being acquired from a number of systems to begin to draw some general conclusions as to the molecular nature of these changes. Alterations in phenotype of the type described above can be divided broadly and imperfectly into two categories.

(i) Phase changes, in which a given characteristic is either expressed or is not. The simplest analogy of the mechanisms used to effect such changes would be an on-off switch.

(ii) Antigenic variation, in which one of a number of alternative phenotypes are expressed. A flawed analogy for this phenomenon would be the expression of a number of informational cassettes that have been withdrawn from a library.

The antigenic variation phenomena described in this chapter are limited to systems that involve intrastrain variability. The evolutionary relationship of such variation to that observed between strains is not always clear. This is largely due to ignorance of the population genetics of bacterial communities. Some characteristics are observed coincidentally with others leading to the concept of clonal descent of many existing strains of bacteria within clinically-important species. For example, six different clones have been discerned among *Escherichia coli* K1 isolates on the basis of the properties of outer membrane proteins and other characters (Achtman *et al.*, 1983). Also, multi-locus enzyme electrophoresis examination has been used to detect allelic variation in structural genes of a variety of different bacterial isolates (Caugant *et al.*, 1985). An analysis of 261 isolates of *E. coli* by these workers indicated that there is considerable genetic diversity amongst isolates bearing the same H, K or O antigens. Although such diversity was less among isolates sharing two or more antigenic determinants, such findings indicate that there is considerable convergence in the evolution of some antigenic determinants. This would be expected perhaps to be greatest for surface structures of bacteria which are perforce exposed to the defence mechanisms of their hosts. Despite the evolutionary constraints that are placed on bacterial surface components there is clearly a substantial advantage for pathogens in being able to alter the antigenic stimulus that they present to the host in order to avoid the defensive effects of the immune system.

Alterations in phenotype involved in phase changes and antigenic variation are frequently dramatic. The genetic mechanisms that mediate these changes are generally reversible within a clone of cells, which would be expected of systems designed for adaptation to changes in the bacterial environment. There are, however, some good examples of irreversible phase changes which are much less easy to accommodate in attributing strategies for survival to these mechanisms. Furthermore much of the direct evidence for phase and antigenic changes comes from laboratory studies where the conditions prevailing in the natural environment may not have been reproduced.

Genetic variation involving gain or loss of extrachromosomal elements

High frequencies of genetic change frequently accompany the carriage by bacteria of extrachromosomal elements such as plasmids or lysogenic phages. Some of the best documented examples of such phenomena include lysogenic conversions of *Corynebacterium diphtheriae* by β phages (Barksdale, 1959; Welkos and Holmes, 1981), of *Salmonella* O antigen types by ϵ phages (Uetake *et al.*, 1955) and changes in colonial morphology and other properties of myobacteria when carrying certain prophages (Juhasz, 1968; Mankiewicz *et al.*, 1969). Variation dependent on gain or loss of plasmids or phages is however of benefit only to populations of bacteria able to receive the particular converting genetic element. The nature of the switch as far as an individual is concerned is off→ON→OFF−on? depending on the vagaries of gene transfer in any particular bacterial population. In most cases the experimental evidence for such variation can only involve observation of the irreversible loss (ON→OFF) part of this very crude genetic switch. The relevance of such gene transfer-dependent phenomena to most individual pathogens in their natural environments is also unknown. For these reasons only passing mention is made to phase and antigenic variations of this type.

Genetic variation involving overt genomic rearrangements

An early indication of the presence of specific genome rearrangements for modulating the expression of virulence determinants and other antigens has been the correlation between changes in observed phenotype and profiles obtained by restriction endonuclease digestion of total genomic DNA. The most convincing method of correlating any specific genome rearrangement with a variation phenomenon is normally first to clone the gene(s) concerned and then to use all or part of the relevant sequences as a probe in a Southern transfer experiment. However, in some instances detectable differences can be seen with unlabelled total DNA provided that appropriate endonuclease digestions are used. For example, *Mycoplasma pulmonis* undergoes a phase change in which opaque (Op^+) colonial variants give rise to transparent (Tr or Op^-) variants at a rate of about 1.2×10^{-8} per colony forming unit (c.f.u.) per generation and with a reversion rate of about 100 times less (Liss and Heiland, 1983). Different and characteristic restriction fragment length polymorphisms for Op^+ and Op^- variants were found after treatment of genomic DNA with, for example, *Dde*I. The precise nature of the genetic rearrangements

Table 1. Genetic switching of phase and antigenic variation.

Type of switch	*Mechanism*	*Example(s)*
ON→OFF→on	Lysogenic conversion or plasmid carriage. Gene transfer reverses switch in population.	Phage conversion in *Salmonella*.
ON→OFF(→on?)	Chromosomal deletion/gene transfer reverses?	Loss of b capsule in *H. influenzae*
$ON_a(OFF_b) \rightarrow ON_b(OFF_a)$	Inversion of controlling element by site-specific recombination system	Phase variation of *Salmonella* flagellar antigens. G segment inversion in phage Mu
ON→OFF ON→OFF→ON	Chromosomal deletion. Deletion/gene conversion. Post-transcriptional regulation.	Pilus phase variation in *N. gonorrhoeae*.
$ON_a \rightarrow (OFF?) \rightarrow ON_n$	Deletion/gene conversion leading to assembly of gene cassettes at expression sites.	Pilus antigenic variation in *N. gonorrhoeae*.

a and b are two alternative states, a . . . n represents one of a number of different states; on represents the hypothetical presence of the on switch arising by gene transfer.

that are implied by these polymorphisms is not yet clear although it has been suggested that lysogeny with a cryptic prophage could be involved (Liss and Heiland, 1983).

The presence of restriction fragment polymorphisms *per se* is not by itself good evidence for the association of specific genome rearrangements with phase or antigenic variation. For example, virulent phase I and avirulent phase II variants of *Coxiella burnetii* differ predominantly in the type of lipopolysaccharide produced generating a phenotypic change that is similar to the smooth-to-rough transition found in Gram-negative bacteria (Hackstadt *et al.*, 1985). A study of such variants revealed that polymorphisms observed following digestion of genomic DNA with *Hae*III were not related to antigenic phase variation but rather to clonal differences (O'Rourke *et al.*, 1985). Some examples of genetic switching mechanisms involving demonstrable genomic rearrangements are shown in *Table 1* and will be discussed below.

Flip-flop (On-Off-On) mechanisms: switching of flagellar antigens in Salmonella and other systems involving DNA inversions

Salmonella typhimurium undergoes a phase change that is manifested by the production of one of two possible antigenic types of flagella. In this case the organism synthesizes either H1 or H2 flagellin and can switch reversibly between the two states (Zieg and Simon, 1980). This switch, which is of the $On_a(Off_b) \leftrightharpoons On_b(Off_a)$ type effects a phase change and antigenic variation simultaneously. The rates of this phase change are of the order of $10^{-3}-10^{-5}$ for some strains of *S. typhimurium* but may be as low as 10^{-7} for strains of *S. abortus-equi* (Iino and Kutsukake, 1980). The molecular mechanism for the switch involves the reversible inversion of a 995-bp sequence of the *S. typhimurium* chromosome that is bounded by 14-bp inverted repeat sequences (Zieg and Simon, 1980) (*Figure 1*). The invertible region carries a promoter that directs transcription into the *H2* gene and an adajcent gene *rh1* (Kutsukake and Iino, 1980; Zieg

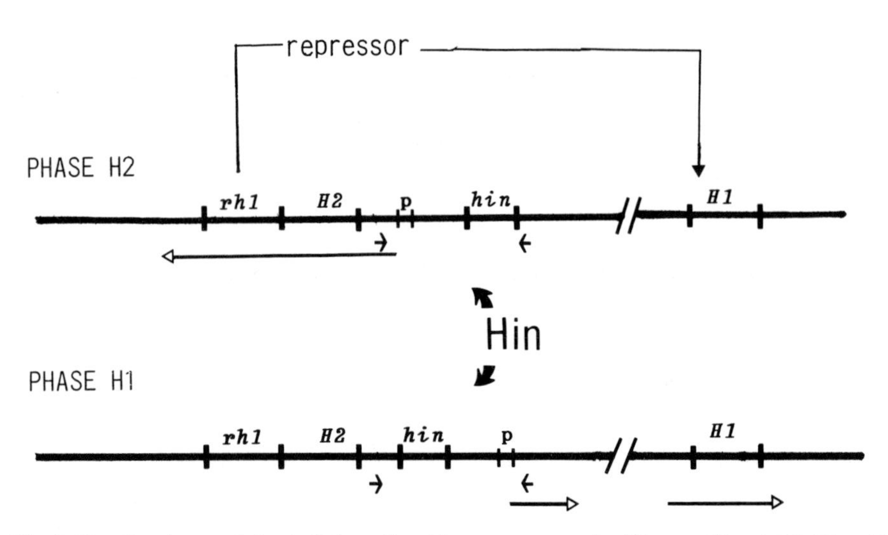

Fig. 1. Flagellar phase variation in *Salmonella*. *rh1*, repressor gene for *H1* gene; *H1* and *H2*, H1 and H2 flagellin structural genes, respectively; short arrows = 14-bp inverted repeats; long arrows indicate direction of transcription. Based on Iino and Kutsukake (1980) and Zieg and Simon (1980).

and Simon, 1980). In one orientation, phase H2, this promoter directs transcription of the *H2* gene leading to synthesis of H2 flagellin and of the *rh1* gene which encodes a *trans*-acting protein repressor that represses expression of the *H1* gene (*Figure 1*). The alternative flagellin gene *H1* is located distantly from the *H2* and invertible control region (Kutsukake and Iino, 1980; Pearce and Stocker, 1967). In the opposite orientation, phase H1, the promoter cannot read through either *H2* or *rh1*. Therefore H2 flagellin is not made and the repression of H1 expression is relieved (*Figure 1*). Inversion of the controlling segment is mediated by a further gene *hin* which is encoded by that segment and is expressed in either orientation (Kutsukake and Iino, 1980; Simon *et al.*, 1980). The *hin* gene product catalyses site-specific recombination between the inverted repeats at either end of the invertible segment and hence causes inversion. The inversion is independent of the normal homologous *recA*-mediated recombination system of the host and resembles many transposition processes.

G region inversion in bacteriophage Mu

A switching event between two alternative expression states similar to that found in *Salmonella* flagellar genes is observed in the invertible 3-kb G segment of the genome of bacteriophage Mu (Kamp *et al.*, 1979; van de Putte *et al.*, 1980). The G segment determines phage host specificity by altering the nature of two gene products encoded by this region of the Mu genome (Giphart-Gassler *et al.*, 1982). Transcription and translation originating from a region that lies outside G through a protein-coding region called S_c (constant) can lead to the production of alternative pairs of proteins (*Figure 2*). In one orientation a protein containing S_c fused to S_v, and a second protein called U are made. These confer on Mu the ability to infect *E. coli*. In the other orientation of G, proteins S_c-S′ and U′ are made which allow the phage to propagate on *Citrobacter freundii* (*Figure 2*). The product of a third gene, *gin* (G inversion), which lies

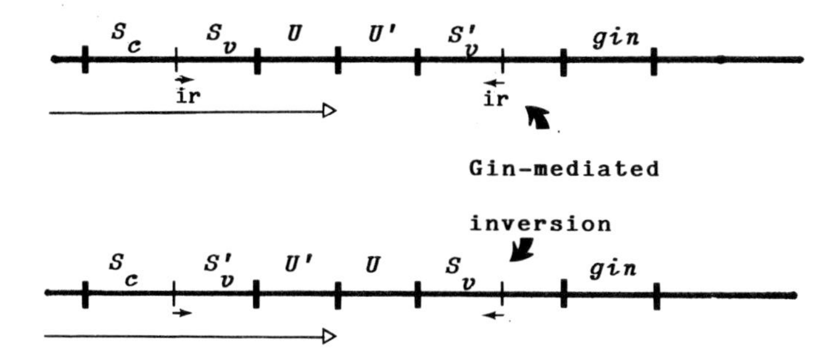

Fig. 2. G-segment inversion in bacteriophage Mu. S_c, constant region of gene *S*; S_v,S'_v, alternative variable regions of gene *S*; *U*,*U'*, alternative segments of gene *U*; *gin*, G inversion invertase gene; other symbols as in *Figure 1*. Based on Giphart-Gassler *et al.* (1982).

outside the G region, is responsible for the site-specific recombination that inverts this controlling segment. Phage P1 has a similar invertible segment and site-specific recombination mechanism that can act interchangeably with that of Mu (Iino and Kutsukake, 1980).

Other systems

The *gin, hin* and *tnp*R (resolvase) genes of the resistance transposon Tn*3* share sequence homologies (Simon *et al.*, 1980). Furthermore, the *gin* gene product can cause site-specific recombination and hence inversion at sites recognized by the *hin* gene product and vice versa. The site-specific recombination system (*cin*) of phage P1 is also capable of inverting the *hin* segment in *Salmonella* (Iino and Kutsukake, 1980) and shows sequence and functional homologies with other invertases (Plasterk and van de Putte, 1984). A further invertible system (*pin*) that complements this group of gene products has been found in a defective prophage-like element called *e14* which is found in the chromosomes of some strains of *E. coli* (Plasterk and van de Putte, 1984; van de Putte *et al.*, 1984). This suggests that there is a family of gene products, probably of common evolutionary origin, involved in genome rearrangements that involve DNA inversions (Plasterk and van de Putte, 1984).

On→Off (→On?) switching systems involving deletion of genetic material. Capsule phase variation in Haemophilus

Encapsulated strains of *Haemophilus influenzae* type b are known to lose readily their ability to produce a capsule. Frequencies of loss of capsule production in this phase change have been estimated to be between 1×10^{-2} and 3×10^{-2} both *in vivo* and *in vitro* (Hosieth *et al.*, 1985). In this case no revertants to the b^+ capsule phenotype were ever isolated, suggesting permanent loss of genetic capacity to manufacture capsular polysaccharide. The most obvious and superficial explanation for these phenomena would be loss of a plasmid that specified all or part of the biosynthetic pathway. However there is no correlation between plasmid carriage and capsular phenotype (Hosieth *et al.*, 1985; E.R.Moxon, personal communication). Similar high frequency loss of capsule production by certain *Klebsiella* capsular serotypes without evident changes in plasmid content has been observed in this laboratory (Allen, Hart and Saunders, in preparation). The change from b^+ to b^- capsule phenotypes in *H. influenzae* was shown by Hosieth *et al.* (1985) to be accompanied by deletion of a 9-kb *Eco*RI fragment of the *H. influenzae* chromosome that had previously been shown to be necessary for expression of b type capsule (Moxon *et al.*, 1984). This 9-kb fragment also hybridizes to DNA from all other *Haemophilus* capsular serotypes suggesting that it encodes some common biosynthetic steps in the production of capsules. The control of b capsule production in *H. influenzae* is therefore controlled by an On→Off genetic switch. One possible advantage of such a mechanism would be avoidance of antibodies against type b capsule. However this would be of no use for the pathogenic potential of the descendants of strains in which the capsular region of the *Haemophilus* genome had been deleted. It has been shown that encapsulated *H. influenzae* strains adhere less well to epithelial cells than non-capsulated variants (Lampe *et al.*, 1982) suggesting the possibility that the b^- form might be preferred for colonization of the nasal or other mucosal surfaces. It is relevant therefore that Hosieth *et al.* (1985) have demonstrated that intranasal inoculation of the rat with b^+ strains can lead to a pronounced shift to the b^- form. However it is difficult to see the long-term advantage that permanent loss of the ability to elicit this capsule would confer on *H. influenzae* since it is a significant virulence determinant. It is possible that *in vivo* b^- (capsule Off) strains could regain the ability to elicit capsule by acquisition of the appropriate region of the *Haemophilus* genome by gene transfer (thus making the switch On→Off→on as far as the *H. influenzae* population as a whole was concerned).

Intragenic recombination and gene conversion: phase and antigenic variation in the pili of Neisseria gonorrhoeae

Neisseria gonorrhoeae is well known for its ability to exhibit interstrain and intrastrain variation in the type of pilus (fimbria) that it produces (see Heckels, Chapter 6). Pili are important virulence factors in gonococci since they mediate adhesion of the bacterium to mucosal surfaces during the initial stages of infection. The genetic potential to vary these major exposed surface structures is one of the reasons for the success of this bacterium in avoiding host defences and in causing repeated infections.

Gonococcal pili are composed of identical pilin subunits of 18 000–22 000 daltons with apparent molecular weight and antigenic differences depending on the strain concerned (Heckels, Chapter 6). Approximately the first 54 amino acids at the amino terminus are conserved in all pili from *N. gonorrhoeae* and constitute the constant (C) region of pilin (Hagblom *et al.*, 1985; Meyer *et al.*, 1984; Rothbard *et al.*, 1985). The remaining 100–110 amino acids extending to the carboxy terminus of the polypeptide are variable and account for the antigenic diversity found in pili. The semivariable (SV) region extends from amino acid residue 54 to 114 and contains various amino acid substitutions. The hypervariable region (HV) which extends from residue 115 to the C terminus contains not only substitutions but also deletions and insertions of 1–4 amino acids in different variant pilins (Hagblom *et al.*, 1985). Immunological and sequence studies show that the HV region is itself flanked by 11- (cys1) and 10- (cys2) residue regions of conserved amino acid sequence that centre around the cysteine residues at positions 121 and 151, respectively (Hagblom *et al.*, 1985; Rothbard *et al.*, 1985)

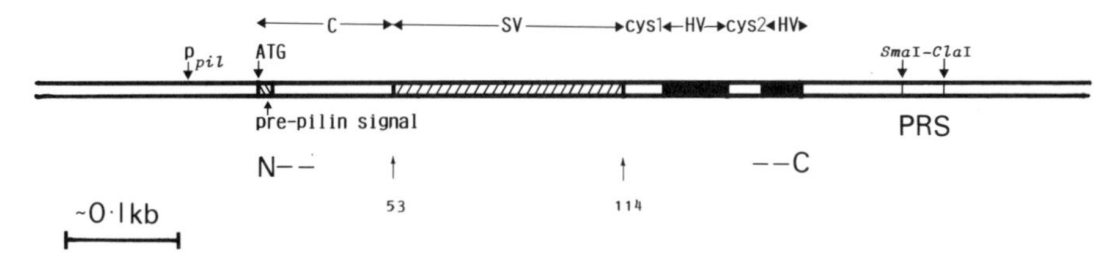

Fig. 3. Genetic arrangement of gonococcal pilus-expressing (*pilE*) sites. p_{pil}, pilin promoter; ATG, translation start codon; C, constant region of pilin; SV, semi variable region of pilin; cys1 and cys2, conserved amino acid sequences around cysteine residues 1 and 2, respectively; HV, hypervariable region of pilin; *Sma*I – *Cla*I (alias PRS) denotes conserved region downstream of pilus-related sequences; PRS, pilus recombination site; N- and -C indicate the amino and carboxy termini, respectively, of mature pilin; numbers beneath arrows indicate approximate amino acid residue positions. Based on data from Hagblom *et al.* (1985), Meyer *et al.* (1984), Segal *et al.* (1986).

(*Figure 3*). The antigenic sites that determine the individual pilus types lie predominantly within the disulphide loop that is formed between these cysteine residues (Heckels, Chapter 6).

Variability in the expression of piliation of *N. gonorrhoeae* involves two separate but interrelated genetic phenomena, a phase change and antigenic variation. In pilus phase variation, pilin genes are switched on (to produce piliated or P^+ cells) or off (to produce non-piliated or P^- cells) at rates which are of the order or $10^{-3}-10^{-5}$. Antigenic variation involves the generation of biochemical and antigenic diversity in pilus type and is caused by the expression of different pilin genes. The two processes may be linked since the transition between P^+ and P^- states is accompanied in some cases (see below) by genome rearrangements. Furthermore P^+ revertants from P^- derivatives may express pili that are antigenically distinct from those produced by the 'parental' P^+ strain (Hagblom *et al.*, 1985; Meyer *et al.*, 1984; Segal *et al.*, 1985).

Phase variation. The ability to alter piliation status by a phase change is of great advantage to the gonococcus because, although pili are important adhesins for initiating infection, they might render the organism more susceptible to host defence mechanisms once access to tissues has been gained. Furthermore, the ability to be non-adherent may assist in the spread of the gonococcus to other sites in the infected host or to other individuals. It is also possible that the genetic mechanism for antigenic variation requires phase variation to the P^- state as an obligatory intermediate step.

Phase variation in the gonococcus is a complex process and the generation of P^- variants from P^+ parental strains seems to occur by genetic switching events of both the ON→OFF→ON and ON→OFF type. Reversible transition between the P^+ and P^- states of *N. gonorrhoeae* MS11 has been reported to be accompanied by genome rearrangements (Meyer *et al.*, 1982; Segal *et al.*, 1985). In contrast, studies on MS11 by Swanson and Koomey (1985) and Swanson *et al.* (1985) have demonstrated that P^- variants may be generated from P^+ strains by several different routes, some of which do not involve demonstrable rearrangement of the gonococcal genome. Furthermore the pilus minus phenotype of the gonococcus, as measured by colony type and electron microscopy, would appear not to be homogeneous (*Table 2*).

Part of the reason for the discrepancy in interpretation of the gonococcal pilus phase

Table 2. Pilus phenotypes of *N. gonorrhoeae*.

Phenotype	*Pili*	*Pilin*	*Pilin-specific mRNA*	*Genomic arrangement*
Piliated				
P^+	+	+(2)	+(2?)	Two non-identical *pilE* genes?
P^+	+	+(1)	+	Two identical *pilE* genes
P^+	+	+	+	One *pilE* gene
Non-piliated, reverting				
P^-	−	+/−	?	Deletions in either *pilE*
P^-rp^-	−	−	+	Single intact *pilE* gene
P^-rp^+	−	+(1 or 2)	+	Single intact *pilE* gene
Non-piliated, non-reverting				
P^-n	−	−	−	Deletion of 5′(C) region(s) of *pilE*
P^-n	−	−	−	Intact *pilE*

Data from Meyer *et al.*, 1984; Nicolson *et al.*, 1986; Segal *et al.*, 1985; Swanson *et al.*, 1985. The notation for pilus minus phenotypes is based on that of Swanson *et al.* (1985).
(1) and (2) indicate the number of distinct pilin polypeptides produced.

change seems to stem from variation that has occurred in the stocks of strain MS11 used in different laboratories (Swanson *et al.*, 1985) and which may therefore behave differently. Lines of MS11 have therefore been given suffixes to identify the laboratory of use and hence avoid confusion (Swanson *et al.*, 1985). Variation seems to occur in the number of pilin expression sites present in different strains and derivatives of such strains. An expression site (*pilE*) contains the complete structural gene for pilin including a promoter and contiguous coding regions for the C, SV and HV regions (*Figure 3*). In addition, each site contains coding DNA for an N-terminal prepilin sequence (Meyer *et al.*, 1984; Segal *et al.*, 1986). This sequence of amino acids is present on the primary translation product of MS11 pilin and is presumably processed prior to the polymerization of pilin molecules into mature pili. The characteristics of the prepilin peptide are not like those of conventional signal sequences and its role is as yet unclear (Meyer *et al.*, 1984). The chromosome of *N. gonorrhoeae* strain $MS11_{ms}$ has been shown to contain two expression loci, *pilE1* and *pilE2*, which are involved in production of pilin and are located within about 20 kb of each other on the chromosome (Meyer *et al.*, 1984; Swanson *et al.*, 1985). When cells of $MS11_{ms}$ are in a piliated state both loci carry intact pilin-coding sequences and their own promoters. Phase variation from P^+ to P^- states is associated with deletion, normally of the 5′ region, of pilus-coding sequences from one or both of the expressing loci (Segal *et al.*, 1985). Such deletions, which vary in size, are believed to involve single or multiple recombination events between directly repeated sequences lying within the *pilE* loci and/or within homologous sequences lying elsewhere. Deletion in only one *pilE* gene can result in little or no pilus expression despite the fact that the remaining gene is apparently intact and should therefore be functional. This has led to the suggestion that there is an additional, *trans*-acting mechanism for regulating pilus expression that could itself be regulated by recombinational switching events or some as yet undefined mechanism (Hagblom *et al.*, 1985; Segal *et al.*, 1985, 1986). Some P^- derivatives observed in this study of MS11 were able to revert to producing pili homologous to those produced by the parental cell line (Segal *et al.*, 1985, 1986). This suggested that the information necessary to produce a specific pilin can be conserved, allowing a functional expressing gene to be regenerated by recombination between a deleted *pilE1* and an intact *pilE2* gene thus causing a P^- to P^+ reversion (Segal *et al.*, 1985).

Studies by Swanson and co-workers (Swanson and Koomey, 1985; Swanson *et al.*, 1985) on other derivatives of MS11 ($MS11_{mk}$ and $MS11_{zm}$) together with a further strain (JS3) show, in contrast, that P^+ to P^- transitions that involve demonstrable genomic rearrangements are invariably non-reverting (P^-n) (*Table 2*). Several different types of pilin gene rearrangement were involved in the generation of P^-n variants but sequences corresponding to the 5′ end of the gene were invariably deleted from the genome. Nevertheless, sequences corresponding to the 3′ end were retained in several copies (Swanson *et al.*, 1985). Furthermore, two classes of reverting non-piliated variants that did not shown genome rearrangements were observed by these workers; P^-rp^- which manufacture pilin-specific mRNA but not pilin and P^-rp^+ which make both the mRNA and pilin but fail to assemble that pilin into mature pili (*Table 2*). These results are consistent with pilus on $\rightleftharpoons$ off switches that operate at the post-transcriptional and/or post-translational levels. Furthermore they are consistent with the proposal advanced by Segal *et al.* (1985) that the P^- state may be determined in some cases, even when *pilE* sites are apparently intact, by an as yet undefined regulator. Reversion rates to P^+ for both P^-rp^+ and P^-rp^- types were of the order of 10^{-2} or 10^{-3} per cell (Swanson *et al.*, 1985). Interestingly, P^-rp^+ variants of strain MS11 were found to elaborate two different pilins of 21 000 and 16 000 which both reacted with anti-pilus antibodies. However, revertants to P^+ produced pilin molecules of 21 000 only. This suggests that the 16 000 pilin molecules may be truncated products that are non-functional and may interfere with pilus assembly by subunit mixing (Swanson *et al.*, 1985). The derivative of strain MS11 used in these studies, $MS11_{mk}$, together with a number of other strains, notably JS3, contains only a single intact copy of the pilin-expressing gene (Swanson *et al.*, 1985). Apparently, therefore, at some stage of the subculturing of MS11 a duplication of a single *pilE* gene or deletion of paired *pil* genes has occurred to generate the observed differences. Whether possession of duplicated *pilE* genes is the norm in gonococci remains to be seen. Strain JS3, which is independent of the MS11 series, certainly only possesses one site (Swanson *et al.*, 1985). Independent studies in this laboratory indicate that only one pilus expression sites is present in the genome of another strain, *N. gonorrhoeae* P9 (Nicolson *et al.*, 1986).

There may actually be good evolutionary reasons for the duplication of *pilE* in the gonococcal genome. A mechanism for phase variation and/or antigenic variation that involves recombination, even if site-specific, would have a tendency to create permanent loss by genomic deletion of the capacity to produce pili. The generation of non-reverting P^-n variants from strain $MS11_{mk}$ which has just one *pilE* copy would tend to support this notion. Loss of ability to produce pili in a strain with duplicated *pilE* genes would probably be obscured in some cases by retention of an intact or repairable copy of the pilin gene. Therefore, all the mechanisms for generating non-piliated variants may not be as apparent in such strains as they are in strains with only one *pilE* locus (Swanson *et al.*, 1985). Possession of more than one expressing site would enable maximization of pilus expression (Saunders, 1985), a not inconsiderable advantage when pilin may constitute 1% or more of the total protein in a P^+ gonococcus (J.Heckels, personal communcation). Expression from more than one expressing site also confers on gonococci the potential ability to elicit two distinct pilins simultaneously (Saunders, 1985). It is clear however that although some strains of the gonococcus contain two expressing sites, one is sufficient for the organism to survive.

Antigenic variation. Pili are substantial surface antigens on *N. gonorrhoeae* and therefore frequent changes in their antigenic nature may assist in the avoidance of the host immune response. Variant pili may also be associated with differential attachment to, and virulence for, animal cells suggesting that one role of antigenic variation may be to permit sequential adhesion to different cell types during the course of natural infections (Heckels, Chapter 6). Single isolates of gonocci can give rise to derivatives, both *in vivo* and *in vitro*, that produce pilins of different apparent molecular weight and with different antigenic properties (Heckels, Chapter 6; Swanson and Barrera, 1983). For example, at least 13 distinct pilin genes can be cloned in *E. coli* from DNA of the well-characterized derivatives of strain P9 (Nicolson *et al.*, 1986). Sequencing of pilin-specific mRNA from derivatives of *N. gonorrhoeae* $MS11_{ms}$ has revealed that this strain can elicit one of at least seven different pili (Hagblom *et al.*, 1985). Swanson and Barrera (1983) have also demonstrated that up to 12 different pilin species may be detected in derivatives of a single gonococcal strain. Sequence studies on strain MS11 indicate that variation in the SV region of the pilin gene is confined to transitions and transversions at two or three positions in a codon to produce amino acid substitutions. Most of the resulting changes in this region involved the introduction of charged amino acids. However, where this occurred there was usually a corresponding second substitution nearby to retain the overall charge distribution of the pilin (Hagblom *et al.*, 1985). Variation in the HV region of MS11 pilin was found to involve single codon amino acid substitutions together with in-frame deletions and insertions of one to four codons. It is apparent from these studies that mature pilin genes can be formed by mixing together different combinations of SV and HV segments with the coding information for the constant N-terminal coding region (Hagblom *et al.*, 1985; Segal *et al.*, 1985, 1986). The large variation in C-terminal sequence that this permits, together with the immunodominance of this part of pilin, explains the low cross-reactivity observed between variant pili isolated from different gonococci. The observed alterations in physicochemical properties of pilins also probably result from the relatively minor alterations in polypeptide length, folding properties and charge that result from the variation process. Studies carried out both *in vivo* and *in vitro* indicate that certain pilus types may be expressed preferentially (Heckels, Chapter 6; Hagblom *et al.*, 1985). This may indicate that there are mechanistic hierarchies in the processes for assembling an intact expressing pilus gene.

In *N. gonorrhoeae* $MS11_{ms}$ the two *pilE* genes apparently encode identical pilin molecules. It is therefore not clear whether these two genes are transcribed simultaneously or alternately *in vivo*. We have shown that *pilE* genes cloned from α, and γ pilus-producing variants of *N. gonorrhoeae* P9 may each be divided into two distinct classes on the basis of the apparent molecular weight and affinity for anti-pilus monoclonal antibodies of pilin encoded, and by differences in the restriction endonuclease cleavage maps of the coding DNA (Nicolson *et al.*, 1986). This indicates that two *pilE* genes of strain P9 could be responsible for the production of distinct translation products.

Furthermore, cell lines derived from single cells of *N. gonorrhoeae* P9 can apparently express one, or other, or both of two antigenically and physically distinct pilins (Perry, Nicolson, Heckels and Saunders, unpublished observations). Some variants of strain MS11 have also been suspected of producing two non-identical pilin messages (M.So, personal communication). This may explain the difficulty encountered in sequencing the 5′ ends of pilin mRNA species from some variants of MS11 (M.So personal communication) and the 3′ ends of others (Hagblom *et al.*, 1985). Whether gonococci can actually simultaneously elicit two distinct pili therefore remains to be seen. However, it is known that some strains of pathogenic *E. coli* contain multiple copies of adhesin gene sequences (Hull *et al.*, 1985) and others can produce three or more physiochemically and antigenically distinct pili at the same time (Karch *et al.*, 1985; G.Boulnois, personal communication).

The chromosome of the gonococcus contains, in addition to one or two *pilE* loci, numerous pilin-related sequences (Meyer *et al.*, 1984; Nicolson *et al.*, 1986; Segal *et al.*, 1986; Swanson *et al.*, 1985). These sequences are apparent when restriction endonuclease digests of gonococcal DNA are hybridized with radiolabelled pilin DNA in Southern transfer experiments (*Figure 4*). The hybridization profiles are consistent with the expressing sites being located at fixed positions on the gonococcal genome and with the silent sequences (*pilS* loci) being scattered at various other locations. The *pilS* loci are partially homologous to the expressed pilin genes but lack their own promoters and represent only part(s) of the DNA sequence found at *pilE* loci (Meyer *et al.*, 1984; Segal *et al.*, 1985, 1986). One such silent locus, *pilS1*, was mapped to within about 15 kb of *pilE1* on the $MS11_{ms}$ genome (Meyer *et al.*, 1984). However, pilin-related sequences are known to be located much closer to and upstream of *pilE* loci in both MS11 (Meyer *et al.*, 1984) and P9 (Nicolson *et al.*, 1986). A region of DNA downstream of the 3′ end of the complete pilin gene has been found to be present at all regions of the MS11 genome (Segal *et al.*, 1986) and most regions of the P9 genome (Nicolson *et al.*, 1986) that contain pilin-related sequences. This region is located about 110 bp downstream of the pilin stop codon and is characterized by the presence of a small *Sma*I – *Cla*I fragment of about 70 bp (*Figure 3*). It has been referred to as the *Sma*I repeat (Meyer *et al.*, 1984; Segal *et al.*, 1985) or more recently PRS1 (Segal *et al.*, 1986) (see below).

The silent regions contain the genetic information necessary for generating antigenic diversity in pili. They may also allow the regeneration of intact pilin-expressing loci during a P^- to P^+ phase change by gene conversion resulting from recombination between a *pilS* locus and a partially deleted *pilE* site (Hagblom *et al.*, 1985). The use of specific synthetic oligonucleotide probes corresponding to the pre-pilin sequence, C, SV, HV and the conserved downstream 3′ sequence of pilin has revealed that the arrangement of silent pilus sequences is complex (Segal *et al.*, 1986). A complete copy of the pilin signal is present only at the expression sites but hybridization data indicate that the DNA coding for the pre-pilin peptide may be split in silent copies of pilus sequences. DNA corresponding to the C region is present in regions upstream of the pilin structural genes in addition to being present at *pilE* sites. This is apparent in P^- variants which do not contain an intact pilin gene but nevertheless hybridize to a C-specific probe (Segal *et al.*, 1986). Similarly, many additional copies of the SV region are present in the genome, including that part upstream of expression sites. In

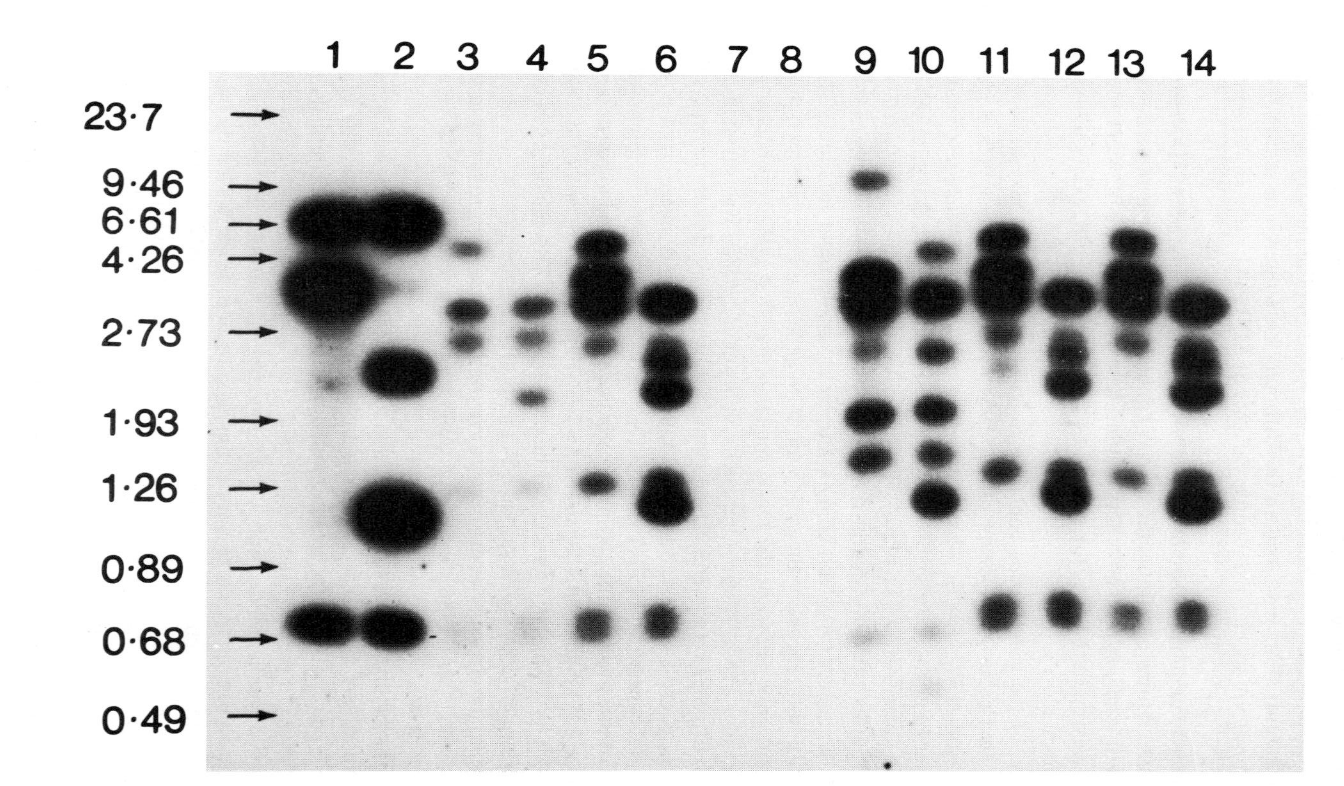

Fig. 4. Pilin-related sequences in pilus variants of *N. gonorrhoeae* P9. Southern transfer hybridization analysis of genomic DNA from pilus variants of *N. gonorrhoeae* P9 was carried out using as a probe a ^{32}P-labelled 1.6-kb *Xba*I–*Pvu*II fragment of P9-2(α) DNA which harbours an intact expressing pilin gene. **Tracks 1:** probe control, plasmid pLV260 digested with *Cla*I; **2:** pLV260 digested with *Cla*I and *Xba*I; **3:** a *Cla*I digest of DNA from a P- variant of P9-2; **4:** as for **3** but *Cla*I–*Xba*I digest; **5:** P9-2(α) DNA digested with *Cla*I; **6:** as for **5** but *Cla*I–*Xba*I digest; **7:** bacteriophage SPP1 DNA digested with *Eco*RI; **8:** bacteriophage λ DNA digested with *Hin*dIII; **9:** P9-20 (β) DNA digested with *Cla*I; **10:** as for **9** but a *Cla*I–*Xba*I digest; **11:** P9-35(γ) DNA digested with *Cla*I; **12:** as for **11** but a *Cla*I–*Xba*I digest; **13:** P9-37(δ) DNA digested with *Cla*I; **14:** as for **13** but a *Cla*I–*Xba*I digest. Arrowed size markers are in kilobases. (I.J.Nicholson and A.Perry, Department of Microbiology, University of Liverpool.)

contrast, hybridization with HV-specific probes indicates that each hypervariable region is carried at only one silent site in addition to being present at *pilE*. The two conserved regions, cys1 and cys2, that lie within the variable region (Hagblom *et al.*, 1985) are present in all pilus-related sequences of MS11 as is the *Sma*I repeat (Segal *et al.*, 1986). These results indicate that silent constant and variable sequences do not reside together in the genome and that they are brought together with the pre-pilin signal sequence only at an expression site (Segal *et al.*, 1986). The presence of the conserved sequences cys1 and cys2 at all pilus-related sequences on the gonococcal genome regardless of whether they are silent copies of the constant 5′ end or variable 3′ end of the pilin gene suggests that these short regions could provide homology for recombination between separated C, SV and HV cassettes (Segal *et al.*, 1986). Furthermore the *Sma*I repeat which lies downstream of the 3′ end of the intact pilin gene is found at silent sequences which include the 5′ constant part of the gene only. For this reason Segal *et al.* (1986) have proposed that this sequence be renamed Pilus Recombination Site 1 (PRS1). They suggest that the role of PRS1 is to promote recombination with like sequences in the *pilS* loci and hence lead to gene conversion at a deleted (or presumably an intact) *pilE* site. PRS sites would thus be analogous to flanking recognition sequences found in the yeast mating type switching system where one of two alternative mating type genes are inserted at an expression (*MAT*) locus (Herskowitz, 1983).

The enzymatic machinery involved in effecting the intragenic recombination events during antigenic variation of gonococcal pili is as yet unknown. It is possible that site-specific recombination enzymes similar to the *hin* or *gin* gene products are necessary. There are so many different rearrangements possible in the pilus-encoding regions of *N. gonorrhoeae* that such a recombination system would have to possess a fairly lax specificity. This could indeed be the explanation for the high incidence of apparently unfavourable rearrangements (e.g. P^+ to P^-n) that occur in gonococci. Alternatively, but *a priori* less likely, would be a battery of enzymatic mechanisms specific for each variant to be generated. A third possibility is that there is sufficient homology between the conserved regions of the participating gene cassettes to allow a normal generalized recombination system (analogous to the *recA* system of *E. coli*) of *N. gonorrhoeae* to be involved. Part of the assembly process at the expressing site could therefore involve simple reciprocal crossing-over events utilizing conserved regions such as cys and PRS1. However, such a mechanism cannot account for all the observed events since, for example large deletions occur that remove all or part of the pilin gene from an expressing site. It is likely therefore that the process of pilus antigenic variation involves gene conversion as well as more simple recombinational events (Hagblom *et al.*, 1985; Segal *et al.*, 1986).

Antigenic variation in the surface antigens of African trypanosomes is superficially similar to that in the gonococcus and has parallels in that several different molecular mechanisms operate to produce the variable phenotype observed (Buck and Eisen, 1985). Variation is achieved by differential activation of specific genes from a large library of antigen-specific sequences and seems to involve both recombinational cross-over and gene conversion mechanisms. Transcription of a specific sequence takes place when it is located at one of a number of alternative expression sites which are always located in telomeres (Van der Ploeg *et al.*, 1984). In *Trypanosoma brucei* activation of antigen genes most frequently involves genome rearrangements that cause gene conversion events. These result from duplicative transposition processes. However telomeric antigen genes can be expressed with little or no rearrangement (see, for example Laurent *et al.*, 1984; Bernards *et al.*, 1984). Furthermore, a third mechanism involving reciprocal recombination upstream of the expression site may be employed, albeit apparently infrequently, to activate an antigen gene. This results in placing the new gene at that site at the expense of the antigen gene previously present. The displaced gene is inactivated by its placing in a silent region of the genome as a consequence of a reciprocal cross-over (Pays *et al.*, 1985). Gene conversion events are also involved in the placing of alternate yeast mating type genes at the *MAT* expression locus in *Saccharomyces cerevisiae* (Klar and Strathern, 1984; Klein, 1984). A major aspect of the situation found both in *T. brucei* and in yeast is that complete silent copies are placed at expression sites whereas in the gonococcus the intact pilin gene is found only at *pilE* with the silent pilin sequences representing just fragments of that gene. However there are distinct similarities between the genetic system used for antigenic variation in gonococci and the generation of diversity in immunoglobulins by recombinational assembly of intact Ig genes (Hagblom *et al.*, 1985; Segal *et al.*, 1986).

Mechanisms that do not involve apparent genome rearrangements

It is to be expected that many examples of phase or antigenic variation will not involve any apparent chromosomal rearrangement. This may be because the rearrangements that occur are too subtle to be detected using available gene probes and restriction endonucleases. Strategies based on examining the genetic location of a gene or genes encoding a particular antigenic determinant may fail to demonstrate rearrangement because unlinked genes may modulate the change in *trans*.

An example of a switching system not involving gross rearrangement is provided by principal outer membrane protein II (PII or Op) of *N. gonorrhoeae* which undergoes a phase change, resulting in a reversible transition between opaque (Op^+) and transparent (Op^- or Tr) colony phenotype at a frequency of up to 10^{-3} per cell (Swanson and Barrera, 1983). In addition, one of a variety of different PII variant proteins with different physicochemical, virulence and antigenic properties can be produced by Op^+ derivatives of a single strain (Swanson, 1982; Swanson and Barrera, 1983). This antigenic variation is analogous to that found in gonococcal pilins. The intrastrain heterogeneity observed in the opacity protein has been shown to correlate with changes found in pili in some cases (Salit *et al.*, 1980) but not in others (Swanson and Barrera, 1983). *opa*, the structural gene for PII, lies about 500 bp downstream of the *pilE1* gene in *N. gonorrhoeae* $MS11_{ms}$ (Stern *et al.*, 1984). Despite this proximity, *pilE1* and *opa* do not appear to be coordinately regulated in all gonococcal strains (Stern *et al.*, 1984; Swanson and Barrera, 1983). The close proximity of expression sites for these two highly variable gene products may indicate that this region of the gonococcal genome is hypermutable. Therefore it would not perhaps be surprising, given the lability of the pilus-switching system, that coincidental changes occur in neighbouring genes. When a cloned *opa* gene was used as a hybridization probe, Stern *et al.* (1984) found that there were a large number of PII-related sequences in the genome. Comparison of the cloned DNA sequences of two different PII variants revealed that there were numerous single base changes and several clustered base differences, including sufficient coding information to produce two amino acid insertions/deletions in the protein sequence (Stern *et al.*, 1984). However, no gross rearrangements of the type associated with pilus phase variation were found by these workers. This may be because any rearrangements involved were too subtle to be detected in the strains examined. It is, therefore, not clear at this stage how the alternating expression of variant protein II species is generated genetically.

The phase variation of type 1 fimbrae of *E. coli* has also been shown to be unaccompanied by significant genome variations, at least in the region of the pilus structural gene (Freitag *et al.*, 1985). In this case however the genetic switch involved is of the on ⇋ off type and does not involve antigenic variation (see below). It may be that reversible inversion of a relatively small region of the genome might go undetected, especially if it were a controlling segment that lies distant from the pilus structural genes.

Other systems

Phase and antigenic variation systems other than those described above have tended to be less well investigated in molecular terms. However, substantial progress is being made in other systems of which antigenic variation in *Borrelia hermsii* (Plasterk, Simon and Barbour, Chapter 8), fimbrial variation in *E. coli* (Smyth, Chapter 7) and phase changes in *Bordetella pertussis* (Robinson, Duggleby, Gorringe and Livey, Chapter 9) are good examples.

Studies on pili other than those of the gonococcus indicate that there may be common mechanisms for modulating phase changes and antigenic variability. Pili isolated from *N. meningitidis* show very strong homologies with the N-terminal constant region of gonococcal pili with only about six amino acid differences in the first 51 amino acids (Hermodsen *et al.*, 1978; Olafson *et al.*, 1985). There is also marked sequence homology between the N-terminal regions of *Neisseria* pilins and pilins isolated from *Bacteroides nodosus* (McKern *et al.*, 1985), *Moraxella nonliquefaciens* (Froholm and Sletten, 1977) and *Pseudomonas aeruginosa* (Sastry *et al.*, 1983). This conserved region of pilins is relatively hydrophobic and may be required for assembly of pili by hydrophobic interactions. Phase variation has been observed for a number of pili including, for example, those of *Moraxella bovis* (Bovre and Froholm, 1972), 987P pili of enterotoxigenic *E. coli* (Nagy *et al.*, 1977) and P-fimbriae of pyelonephritic *E. coli* (Rhen *et al.*, 1983; Nowicki *et al.*, 1984).

Both inter- and intrastrain antigenic variation has been observed frequently in pili from different bacteria. For example, meningococci exhibit considerable heterogeneity in the type of pili produced with respect to both apparent molecular weight of the pilin subunit and antigenic specificity (Diaz *et al.*, 1984; Olafson *et al.*, 1985). A comparison of the pili isolated from serogroups A and E of *B. nodosus* has shown that the first 44 amino acid residues at the N terminus are the same but that about a third of the remaining 107 C-terminal residues are different (McKern *et al.*, 1985). DNA sequence studies have also revealed that there are about 30 amino acid differences between the major serological variants of K88 pili (K88ab, K88ac and K88ad) of pathogenic *E. coli* (Josephsen *et al.*, 1985). Variable regions rather like those in gonococcal pili are found between amino acid residues 162 and 175 in K88 pilus protein. In K88ac this region contains a deletion of three amino acid residues together with a single amino acid insertion between residues 104 and 105 (Josephsen *et al.*, 1985). This pattern of amino acid substitution, deletion and insertion is apparently similar to that on gonococcal pilins. However it is not yet clear whether the resulting alterations in protein structure have any consequence in mediating intra- and interstrain antigenic variation.

A potential capacity for variation may first become apparent by demonstration that the coding region for a particular antigen is present as more than one intact or partial copy of the genome of the producing bacterium. For example, the *snp5* gene which encodes the fibrillar serotype 5 M protein of group A streptococci is present in multiple copies in the streptococcal genome (Kehoe *et al.*, 1985). Two of the copies can be expressed in *E. coli* indicating that they are presumably intact in the streptococcus and probably encode very similar polypeptides. It is not yet known whether these and other partially homologous sequences are expressed in streptococci or whether they are capable of encoding antigenically variant 5 M proteins. The multiple *snp5* copies might therefore play the same role as the different pilin-related sequences and expression loci in gonococci. However, over 70 different serotypes of 5 M protein are known but only one is normally expressed in each strain.

Mutation/selection or environmental signalling

There are two likely methods by which the environment might interact with the genetic apparatus for mediating variation in the antigenic components of bacteria. Firstly, genetic instabilities and other mutational mechanisms may simply generate random changes in particular genes. Selection pressures exerted by the external environment would then act to favour or ensure survival of only one of the phenotypes. The alternative mechanism for modulating the changes would be an environmental signal specific for the variable gene(s) concerned. The simplest target for such regulation would be transcription of the structural gene(s) for the structural component concerned. It has been proposed that type 1 piliation of *E. coli* is under a transcriptional control which may be influenced by cultural conditions (Eisenstein, 1981). A protein of 23 000 encoded by the *hyp* gene of *E. coli* acts in *trans* to regulate the level of piliation (Orndorff and Falkow, 1984a, 1984b). A second separate *trans*-acting factor which is permissive rather than repressive and a *cis*-acting site have been discovered by Freitag *et al.* (1985). Either *trans*-acting factor might fulfil the role of a transcriptional or other regulator that is activated by environmental conditions. It is possible that the *cis*-acting factor found by Freitag *et al.* (1985) represents a recognition site for an invertase (possibly the *trans*-acting function found by these workers would fulfil this role). It has also been shown that transcription of the genes for the virulence pili of some pathogenic *E. coli* is controlled by growth temperature (Goransson and Uhlin, 1984). This indicates that environmental conditions can modulate the expression of at least some surface antigens by a direct genetic mechanism.

Phase variation in *Bordetella pertussis* and *B. bronchiseptica* involves the simultaneous loss of a variety of putative virulence determinants, including toxins, haemolysins and agglutinins (Goldman *et al.*, 1984; Robinson, Duggleby, Gorringe and Livey, Chapter 9; Weiss and Falkow, 1984). The transition from virulent to avirulent phases occurs readily *in vitro*. The reverse phase change to virulence occurs very infrequently *in vitro* in *B. pertussis* (Weiss and Falkow, 1984). A second type of variation, termed antigenic modulation, involving similar changes has been observed in *Bordetella* but in this case the changes can be reversed by alteration of the growth medium (Ezzell *et al.*, 1981). The frequency of phase variation *in vitro* in *B. bronchiseptica* is about 10^{-6} per cell per generation (Lax, 1985). Lax (1985) has therefore proposed that this variation results from random mutations in a controlling region followed by clonal selection. A contrasting view that the observed variations result from an environmental signal, possibly triggering the inversion of a regulatory segment of the *Bordetella* genome, has been proposed by Weiss and Falkow (1984). Current progress in this area of *Bordetella* variation is reviewed by Robinson, Duggleby, Gorringe and Livey (Chapter 9). At present there is little conclusive experimental evidence to distinguish between selective and signalling events in most of the systems that have been studied. It is quite likely that the variable phenotypes observed arise as a consequence of a combination of both processes in many systems. Moreover, it would be unlikely that any selection or signalling system would be the same in different pathogens. A major problem remains in reproducing the authentic environmental selection/signalling conditions that exist *in vivo* whilst retaining the ability to analyse the underlying genetic mechanisms involved.

Conclusions

It is clear that a variety of mechanisms are responsible for the ability of many pathogenic bacteria to alter their antigenic status. Genetic studies on a relatively small number of variations have revealed that genomic rearrangements and specific *trans*-acting protein regulators are frequently involved. However, other mechanisms may well be operating because the vast majority of phase and antigenic variation phenomena have remained uninvestigated by the techniques of molecular genetics. Where investigations have been carried out there has been an inevitable tendency to concentrate on systems that confer a readily identifiable phenotype on variant bacteria. Many of these may be inherently unstable genetically and may not even represent the true behaviour of bacteria in their natural environments. The extent of real or potential antigenic variation in bacteria is however likely to become more apparent as detailed molecular and immunological analyses are carried out on increasing numbers of specific antigens.

Acknowledgements

P.Allen, C.A.Hart, J.E.Heckels, I.J.Nicolson, A.Perry, M.So and J.Swanson are thanked for providing information prior to publication. Part of the work described in this chapter was supported by grants from the Medical Research Council and The Wellcome Trust.

References

Achtman,M., Mercer,A., Kusecek,B., Pohl,A., Heuzenroeder,M., Aaronson,W., Sutton,A. and Silver,R.P. (1983) Six widespread bacterial clones among *Escherichia coli* K1 isolates. *Infect. Immun.*, **39**, 315–335.

Barksdale,L. (1959) Lysogenic conversions in bacteria. *Bacteriol. Rev.*, **23**, 202–212.

Bernards,A., De Lange,T., Michels,P., Liu,A., Huisman,M. and Borst,P. (1984) Two modes of activation of a single surface antigen gene in *Trypanosoma brucei*. *Cell*, **36**, 163–170.

Bovre,L. and Froholm,L.O. (1972) Variation of colony morphology reflecting fimbriation in *Moraxella bovis* and two reference strains of *M. nonliquefaciens*. *Acta Pathol. Microbiol. Scand. Sect. B*, **80**, 629–640.

Buck,G.A. and Eisen,H. (1985) Regulation of the genes encoding variable surface antigens in African trypanosomes. *Am. Soc. Microbiol. News,* **51**, 118–122.

Caugant,D.A., Levin,B.R., Orskov,I., Orskov,F., Svanborg-Eden,C. and Selander,R.K. (1985) Genetic diversity in relation to serotype in *Escherichia coli. Infect. Immun.*, **49**, 407–413.

Diaz,J.L., Virji,M. and Heckels,J.E. (1984) Structural and antigenic differences between two types of meningococcal pili. *FEMS Microbiol. Lett.*, **21**, 181–184.

Eisenstein,B.I. (1981) Phase variation of type I fimbriae in *Escherichia coli* is under transcriptional control. *Science,* **214**, 337–339.

Essell,J.W., Dobrogosz,W.J., Kloos,W.E. and Manclark,C.R. (1981) Phase-shift markers in *Bordetella*: alterations in envelope proteins. *J. Infect. Dis.*, **143**, 562–569.

Freitag,C.S., Abraham,J.M., Clements,J.R. and Eisenstein,B.I. (1985) Genetic analysis of the phase variation control of expression of type 1 fimbriae in *Escherichia coli. J. Bacteriol.*, **162**, 668–675.

Froholm,L.O. and Sletten,K. (1977) Purification and N-terminal sequence of a fimbrial protein from *Moraxella nonliquefaciens. FEBS Lett.*, **73**, 29–32.

Giphart-Gassler,M., Plasterk,R.H.A. and van de Putte,P. (1982) G inversion in bacteriophage Mu: a novel way of gene splicing. *Nature,* **297**, 339–342.

Goldman,S., Hanski,E. and Fish,F. (1984) Spontaneous phase variation in *Bordetella pertussis* is a multistep non-random process. *EMBO J.*, **3**, 1353–1356.

Goransson,M. and Uhlin,B.E. (1984) Environmental temperature regulates transcription of a virulence pili operon in *Escherichia coli. EMBO J.*, **3**, 2885–2888.

Hackstadt,T., Peacock,M.G., Hitchcock,P.J. and Cole,R.L. (1985) Lipopolysaccharide variation in *Coxiella burnetii*: intrastrain heterogeneity in structure and antigenicity. *Infect. Immun.*, **48**, 359–365.

Hagblom,P., Segal,E., Billyard,E. and So,M. (1985) Intragenic recombination leads to pilus antigenic variation in *Neisseria gonorrhoeae. Nature,* **315**, 156–158.

Hermodsen,M.A., Chen,K.C.S. and Buchanan,T.M. (1978) *Neisseria* pili proteins: amino-terminal amino acid sequences and identification of an unusual amino acid. *Biochemistry,* **17**, 442–445.

Herskowitz,I. (1983) Cellular differentiation, cell lineages and transposable genetic cassettes in yeast. *Curr. Top. Dev. Biol.*, **18**, 1–14.

Hosieth,S.K., Connelly,C.J. and Moxon,E.R. (1985) Genetics of spontaneous, high-frequency loss of b capsule expression in *Haemophilus influenzae. Infect. Immun.*, **49**, 389–395.

Hull,S., Clegg,S., Svanborg-Eden,C. and Hull,R. (1985) Multiple forms of genes in pyelonephritogenic *Escherichia coli. Infect. Immun.*, **47**, 80–83.

Iino,T. and Kutsukake,K. (1980) *Trans*-acting genes of bacteriophages P1 and Mu mediate inversion of a specific DNA segment involved in flagellar phase variation of *Salmonella. Cold Spring Harbor Symp. Quant. Biol.*, **45**, 11–16.

Josephsen,J., Hansen,F., de Graaf,F.K. and Gaastra,W. (1985) The nucleotide sequence of the protein subunit of the K88ac fimbriae of porcine enterotoxigenic *Escherichia coli. FEMS Microbiol. Lett.*, **25**, 301–306.

Juhasz,S.E. (1968) Growth retardation, colonial changes and nutritional deficiency in *Mycobacterium phlei* due to lysogeny. *J. Gen. Microbiol.*, **52**, 237–241.

Kamp,D., Chow,L.T., Broker,T.R., Kwol,D., Zipser,D. and Kahmann,R. (1979) Site-specific recombination in phage Mu. *Cold Spring Harbor Symp. Quant. Biol.*, **43**, 1159–1167.

Karch,H., Leying,H., Buscher,K.-H., Kroll,H.-P. and Opferkuch,W. (1985) Isolation and separation of physicochemically distinct fimbrial types expressed on a single culture of *Escherichia coli* 07:K1:H6. *Infect. Immun.*, **47**, 549–554.

Kehoe,M.A., Poirier,T.P., Beachey,E.H. and Timmis,K.N. (1985) Cloning and genetic analysis of serotype 5M protein determinant of group A streptococci: evidence for multiple copies of the M5 determinant in the *Streptocococcus pyogenes* genome. *Infect. Immun.*, **48**, 190–197.

Klar,A.J.S. and Strathern,J.N. (1984) Resolution of recombination intermediates generated during yeast mating type switching. *Nature,* **310**, 744–748.

Klein,H.L. (1984) Lack of association between intrachromosomal gene conversion and reciprocal exhange. *Nature,* **310**, 748–753.

Kutsukake,K. and Iino,T. (1980) A *trans*-acting factor mediates inversion of a specific DNA segment in flagellar phase variation of *Salmonella. Nature,* **284**, 479–481.

Lampe,R.M., Mason,E.O., Kaplan,S.L., Umstead,C.L., Yow,M.D. and Feigin,R.D. (1982) Adherence of *Haemophilus influenzae* to buccal epithelial cells. *Infect. Immun.*, **35**, 166–172.

Laurent,M., Pays,E., Delinte,K., Magnus,E., Van Miervenne,N. and Steinert,M. (1984) Evolution of a trypanosome surface antigen repertoire linked to non-duplicative gene activation. *Nature,* **308**, 370–373.

Lax,A.J. (1985) Is phase variation in *Bordetella pertussis* caused by mutation and selection? *J. Gen. Microbiol.*, **131**, 913–917.
Liss,A. and Heiland,R.A. (1983) Colonial opacity variation in *Mycoplasma pulmonis. Infect. Immun.*, **41**, 1245–1251.
Mankiewicz,E., Liivak,M. and Dernuet,S. (1969) Lysogenic mycobacteria: phase variations and changes in host cells. *J. Gen. Microbiol.*, **55**, 409–416.
KcKern,N.M., O'Donnell,I.J., Stewart,D.J. and Clark,B.L. (1985) Primary structure of pilin protein from *Bacteroides nodosus* strain 216: comparison with the corresponding protein from strain 198. *J. Gen. Microbiol.*, **131**, 1–6.
Meyer,T.F., Mlawer,N. and So,M. (1992) Pilus expression in *Neisseria gonorrhoeae* involves chromosomal rearrangement. *Cell*, **30**, 45–52.
Meyer,T., Billyard,E., Haas,R., Strozbach,S. and So,M. (1984) Pilus genes of *Neisseria gonorrhoeae*: chromosomal organization and DNA sequence. *Proc. Natl. Acad. Sci. USA*, **81**, 6110–6114.
Moxon,E.R., Deich,R.A. and Connelly,C. (1984) Cloning of chromosomal DNA from *Haemophilus influenzae*. Its use for studying the expression of type b capsule and virulence. *J. Clin. Invest.*, **73**, 298–306.
Nagy,B., Moon,H.W. and Isaacson,R.E. (1977) Colonization of porcine intestine by enterotoxigenic *Escherichia coli*: selection of piliated forms *in vivo*, adhesion of piliated forms to epithelial cells *in vitro* and incidence of a pilus antigen among porcine enteropathogenic *E. coli. Infect. Immun.*, **16**, 344–352.
Nicolson,I.J., Perry,A.C.F., Heckels,J.E. and Saunders,J.R. (1986) Genetic analysis of variant pilin genes from *Neisseria gonorrhoeae* P9 cloned in *Escherichia coli*: physical and immunological properties of encoded pilins. *J. Gen. Microbiol.*, submitted.
Nowicki,B., Rhen,M., Vaisanen-Rhen,V., Pere,A. and Korhonen,T.H. (1984) Immunofluorescence study of fimbrial phase variation in *Escherichia coli* KS71. *J. Bacteriol.*, **160**, 691–695.
Olafson,R.W., McCarthy,P.J., Bhatti,A.R., Dooley,J.S.G., Heckels,J.E. and Trust,T.J. (1985) Structural and antigenic analysis of meningococcal pili. *Infect. Immun.*, **48**, 336–342.
Orndorff,P.E. and Falkow,S. (1984a) Organization and expression of genes specifying type 1 piliation in *Escherichia coli. J. Bacteriol.*, **159**, 736–744.
Orndorff,P.E. and Falkow,S. (1984b) Identification and characterization of a gene product that regulates type 1 piliation. *J. Bacteriol.*, **160**, 61–66.
O'Rourke,A.T., Peacock,M., Samuel,J.E., Frazier,M.E., Nayvig,D.O., Mallavia,L.P. and Baca,O. (1985) Genomic analysis of phase I and phase II *Coxiella burnetii* with restriction endonucleases. *J. Gen. Microbiol.*, **131**, 1543–1546.
Pays,E., Guyaux,M., Aerts,D., Van Meirvenne,N. and Steinert,M. (1985) Telomeric reciprocal recombination as a possible mechanism for antigenic variation in trypanosomes. *Nature*, **316**, 562–564.
Pearce,U.B. and Stocker,B.A.D. (1967) Phase variation of flagellar antigens in *Salmonella*: abortive transduction studies. *J. Gen. Microbiol.*, **49**, 335–349.
Plasterk,R.H.A. and van de Putte,P. (1984) Genetic switches by DNA inversions in prokaryotes. *Biochim. Biophys. Acta*, **782**, 111–119.
Rhen,M., Makela,P.H. and Korhonen,T.H. (1983) P-fimbriae of *Escherichia coli* are subject to phase variation. *FEMS Microbiol. Lett.*, **19**, 267–271.
Rothbard,J.B., Fernandez,R., Wang,L., Teng,N.N.H. and Schoolnik,G.K. (1985) Antibodies to peptides corresponding to a conserved sequence of gonococcal pilins block bacterial adhesion. *Proc. Natl. Acad. Sci. USA*, **82**, 915–919.
Salit,I.E., Blake,M. and Gotschlich,E.C. (1980) Intra-strain heterogeneity of gonococcal pili is related to opacity colony variance. *J. Exp. Med.*, **151**, 716–725.
Sastry,P.A., Pearlstone,J.R., Smillie,L.B. and Paranchych,W. (1983) Amino acid sequence of pilin isolated from *Pseudomonas aeruginosa. FEBS Lett.*, **151**, 253–256.
Saunders,J.R. (1985) Antigenic variation in gonococcal pili explained. *Nature*, **315**, 100–101.
Segal,E., Billyard,E., So,M., Strobach,S. and Meyer,T. (1985) Role of chromosomal rearrangement in *N. gonorrhoeae* pilus phase variation. *Cell*, **40**, 293–300.
Segal,E., Hagblom,P., Seifert,H.S. and So,M. (1986) Antigenic variation of gonococcal pilus results from assembly of separated silent gene segments. *Cell*, in press.
Simon,M., Zieg,J., Silverman,M., Mandel,G. and Doolittle,R. (1980) Phase variation: evolution of a controlling element. *Science*, **209**, 1370–1374.
Stern,A., Nickel,P., Meyer,T. and So,M. (1984) Opacity determinants of *Neisseria gonorrhoeae* and chromosomal linkage to the gonococcal pilus gene. *Cell*, **37**, 447–456.
Swanson,J. (1982) Colony opacity and protein II composition of gonococci. *Infect. Immun.*, **37**, 359–368.

Swanson,J. and Barrera,O. (1983) Gonococcal pilus subunit size heterogeneity correlates with transition in colony piliation phenotype, not with changes in colony opacity. *J. Exp. Med.*, **158**, 1459–1468.
Swanson,J. and Koomey,M. (1985) Do changes in pilus expression for *Neisseria gonorrhoeae* involve chromosomal rearrangement. *UCLA Symp. Mol. Cell. Biol., New Ser.*, **20**, 137–152.
Swanson,J., Bergstrom,S., Barrera,O., Robbins,K. and Corwin,D. (1985) Pilus$^-$ gonococcal variants. Evidence for multiple forms of piliation control. *J. Exp. Med.*, **162**, 729–744.
Uetake,H.T., Nagakawa,T. and Akiba,T. (1955) The relationship of bacteriophage to antigenic changes in group E salmonellas. *J. Bacteriol.*, **69**, 571–579.
Van der Ploeg,L.H.T., Schwartz,D.C., Cantor,C.R. and Borst,P. (1984) Antigenic variation in *Trypanosoma brucei* analysed by electrophoretic separation of chromosome-sized DNA molecules. *Cell*, **37**, 77–84.
van de Putte,P., Cramer,S. and Giphart-Gassler,M. (1980) Invertible DNA segment controls host range specificity of bacteriophage Mu. *Nature*, **286**, 218–222.
van de Putte,P., Plasterk,R. and Kuijpers,A. (1984) A Mu *gin* complementing function and an invertible DNA region in *Escherichia coli* K-12 are situated on the genetic element *e*14. *J. Bacteriol.*, **158**, 517–522.
Weiss,A.A. and Falkow,S. (1984) Genetic analysis of phase changes in *Bordetella pertussis*. *Infect. Immun.*, **43**, 263–269.
Welkos,S.L. and Holmes,R.K. (1981) Regulation of toxigenesis in *Corynebacterium diphtheriae* II. Genetic mapping of a tox regulatory mutation in bacteriophage beta. *J. Virol.*, **37**, 946–954.
Zieg,J. and Simon,N. (1980) Analysis of the nucleotide sequence of an invertible controlling element. *Proc. Natl. Acad. Sci. USA*, **77**, 4196–4200.

CHAPTER 6

Gonococcal antigenic variation and pathogenesis

J.E. HECKELS

Department of Microbiology, Southampton University Medical School, Southampton General Hospital, Southampton, UK

Introduction

Gonococcal infections continue to pose worldwide health care problems, and gonorrhoea control has become complicated in several areas by the high incidence of β-lactamase-producing strains. The possibility of developing an effective gonococcal vaccine is therefore an attractive alternative and has stimulated research into the pathogenesis of gonorrhoea and the immunobiology of the gonococcus.

Symptoms of gonorrhoea result from the destruction of the urethral mucosa and the accumulation of gonococci and polymorphonuclear leukocytes (PMN) in subepithelial connective tissue (Harkness, 1948). The first stage in the pathogenic process is adhesion of gonococci to non-ciliated columnar epithelial cells, enabling the bacteria to become established despite the fluid flows to which they are subjected. Some organisms are then engulfed by the host cell membrane and lie within a phagocytic vacuole within the cell (Ward *et al.*, 1974). Intracellular multiplication occurs and bacteria may then penetrate into the subepithelial connective tissue giving rise to the inflammatory response for the typical symptoms of gonorrhoea. Alternatively, gonococci may avoid normal host defences so leading to the complications of disseminated infections. The ability of gonococci to invade the different mucosal surfaces of the human genital tract distinguishes them from commensal neisseria and other normal flora bacteria.

Much work has been concentrated on the major gonococcal surface antigens since they are involved in those gonococcal–host interactions such as colonization of the mucosal surfaces of the genital tract, invasion of epithelial cells and resistance to host defences which determine the eventual outcome of an infection. Such studies have been complicated by the discovery that not only does considerable antigenic diversity exist between strains, but also that a single strain may undergo antigenic shift to produce immunologically distinct surface proteins. This chapter will describe the immunochemical basis of antigenic shift and its potential effect on the pathogenesis of gonococcal infections.

Gonococcal surface antigens

Pili and their association with virulence

Pili, hair-like filamentous appendages, which extend several microns from the bacterial surface, are invariably present on freshly isolated gonococci (Pil$^+$) but not after unselected subculture *in vitro* (Pil$^-$). The importance of pili in virulence was indicated by the observation that loss of virulence on laboratory subculture (Kellog *et al.*, 1963) was concomitant with the loss of pilus expression (Swanson *et al.*, 1971). Many studies have subsequently shown that pili facilitate adhesion of gonococci to a range of human cells including tissue culture cells (Swanson, 1973), vaginal epithelial cells (Mardh and Westrom, 1976), fallopian tube epithelium (Ward *et al.*, 1974) and buccal epithelial cells (Punsalang and Sawyer, 1973). These studies have led to the view that the role of pili in virulence is associated with their ability to promote adherence to mucosal surfaces of the genital tract. One model is that attachment is a two-stage process. Initially pili are able to overcome the electrostatic repulsive barrier which exists between the negatively charged surfaces of gonococcus and host cell (Heckels *et al.*, 1976). This increases the probability of a closer approach leading to a stable adhesion which involves the gonococcal outer membrane and host cell surface.

Outer membrane composition

Outer membranes were first isolated from gonococci by fractionation of membrane preparations using density gradient centrifugation (Johnston and Gotschlich, 1974) but this technique has proved unsatisfactory for isolation of large amounts of material. Subsequent studies have generally utilized the outer membrane vesicles which are released when gonococci are shaken with lithium salts at 45°C (Johnston *et al.*, 1976; Heckels, 1977) or have treated a crude envelope preparation with sodium lauryl sarcosinate to remove cytoplasmic membrane contamination (Walstad *et al.*, 1977). Analysis of such preparations showed that outer membranes of gonococci like those of other Gram-negative bacteria contain lipopolysaccharide and a limited number of proteins (*Figure 1*).

One protein, which can account for over 60% of the total membrane protein, has been termed the major or principle outer membrane protein and is now designated pro-

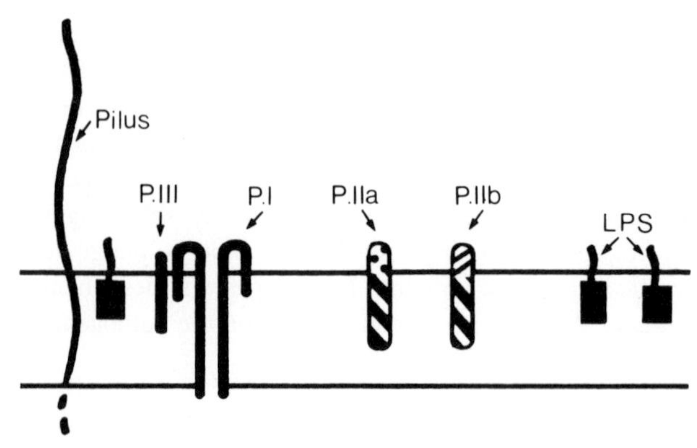

Fig. 1. Major components of the gonococcal outer membrane showing aqueous pore formed by protein I and conserved and variable regions of protein II.

tein I (P.I) (Heckels, 1977; Swanson and Heckels, 1980). P.I varies between different strains with a molecular weight in the range 32 000 (32 K) to 40 K and is a major determinant of serotype specificity (Johnston *et al.*, 1976; Sandstrom *et al.*, 1982). P.I is surface exposed (Heckels, 1978), exists in the membrane as a trimer (Newhall *et al.*, 1980) and can also be linked to the underlying peptidoglycan (Heckels, 1979). These properties are consistent with the observation that P.I functions as a porin creating a hydrophilic channel in the outer membrane (Douglas *et al.*, 1981; Lynch *et al.*, 1984). P.I can be transferred from the outer membrane into the membrane of host cells, a mechanism which might be related to the ability of gonococci to penetrate epithelial cells (Blake and Gotschlich, 1983).

In addition to P.I, initial studies identified a second class of outer membrane proteins (protein II, P.II) which were present in most, though not all, isolates examined and showed characteristic heat modification on sodium dodecyl sulphate–polyacrylamide gel electrophoresis (SDS–PAGE) (Heckels, 1977). P.II, like P.I, is exposed on the gonococcal surface but it does not span the membrane. P.II has been implicated in a variety of host interactions and is subject to antigenic variation within a strain as discussed in detail below. The third major protein in outer membranes, protein III, occurs in close association with P.I (McDade and Johnson, 1980). It is a protein of 31 K molecular weight and appears similar, if not identical, in all strains (Judd, 1982). No important virulence functions have yet been ascribed to it.

Antigenic variation

In vitro studies

Early studies by Kellog and co-workers demonstrated that a single strain of gonococcus growing on solid medium gave rise to variants with characteristically different colonial morphology, designated types (T1, T2, T3 and T4) (Kellog *et al.*, 1963). Primary isolates produced T1 and T2 colonies whereas T3 and T4 predominated after repeated laboratory subculture, although each type could be stably maintained by careful colonial selection during subculture. The T1, T2→T3, T4 switch was shown to be accompanied by loss of pilus expression (Swanson *et al.*, 1971; Jephcott *et al.*, 1971). Subsequently two independent reports correlated the difference between types T3 and T4 with alteration in expression of outer membrane proteins other than P.I (Walstad *et al.*, 1977; Swanson, 1978). These findings suggests an explanation for the observation that expression of outer membrane P.II varied between cultures of the strain P9 (Heckels, 1977, 1978). Using an improved colony typing system we were able to select a series of variants of strain P9 which differed in colonial opacity (Lambden and Heckels, 1979). Surface labelling of intact organisms, or SDS–PAGE of purified outer membranes, showed that the variant with transparent colonial phenotype (O^-) did not express P.II. In contrast a series of opaque colonial variants (O^+) were obtained which produced one or two P.II species (P.IIa, P.IIb, etc.) in the molecular weight range 27–29 K (*Figure 2*). Subsequently P9 variants expressing further P.II types were obtained after application of an *in vivo* selection pressure by growth of gonococci in subcutaneous chambers implanted in guinea-pigs (McBride *et al.*, 1981). Studies with other strains (Black *et al.*, 1984; Swanson, 1982) have confirmed that gonococci generally have the ability to produce up to seven or more distinct P.II species and that these may

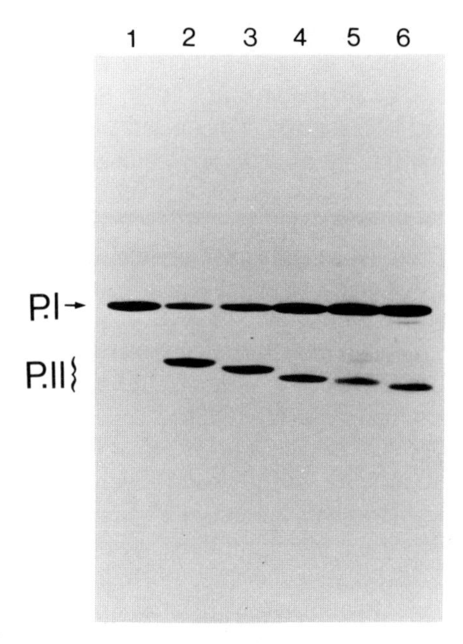

Fig. 2. Variation in outer membrane protein II in colonial variants of *N. gonorrhoeae* strain P9. Outer membranes were prepared from a series of different colonial types and subjected to SDS–PAGE. **Track 1**, transparent colonial variant lacking P.II; **tracks 2–6**, opaque variants producing P.II; species P.IId, P.IIa, P.IIb, P.IIe, P.IIc with molecular weights of 28.9, 28.5, 28, 27.8 and 27.5 K, respectively.

occur in a variety of combinations. Most variants express one or two P.II species but occasionally up to four have been detected in a single variant (Swanson and Barrera, 1983a).

Following discovery of the relationship between opacity and P.II expression, pili were purified from opaque and transparent colonial forms of strain P9 (Lambden *et al.*, 1980). Pili from the transparent type had a pilin subunit molecular weight of 19.5 K (α pili) whereas those from the opaque type were 20.5 K (β pili) (*Figure 3*). Another study with a series of strains showed that pilin molecular weight usually, although not always, varied between similar pairs of opacity variants (Salit *et al.*, 1980). These observations suggested that alteration in pilin molecular weight might be linked to opacity and hence P.II expression. However, further studies with strain P9 identified two further pilus types, γ and δ (*Figure 3*), (Lambden *et al.*, 1981) and showed that each type could be produced by both O^+ and O^- colonial forms (Lambden, 1982). Similarly, two variants of another strain, with no apparent difference in colonial morphology, expressed pili with a distinct difference in molecular weight (Perera *et al.*, 1982). The conclusion that the apparent connection between colonial opacity and pilus expression was fortuitous was confirmed with a study of other strains, which also suggested that a single strain might have the ability to produce at least a dozen different molecular forms of pilin (Swanson and Barrera, 1983b).

By selection of gonococcal variants during laboratory culture it is theoretically possible to isolate a large number of variants which differ in expression of both pili and P.II. Thus we have so far isolated over 30 different variants of strain P9 which have been

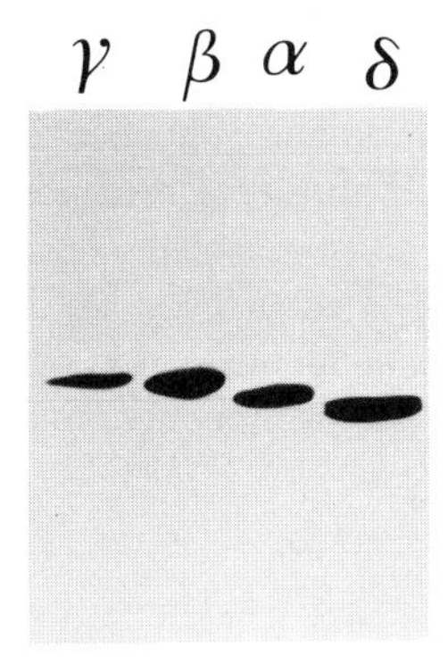

Fig. 3. Variation in subunit molecular weight of pili from colonial variants of *N. gonorrhoeae* strain P9. Pili were purified from $P.II^+$ and $P.II^-$ variants and subjected to SDS–PAGE; γ pili, molecular weight 21 K; β pili, 20.5 K; α pili, 19.5 K and δ pili, 18.5 K.

utilized in studies designed to investigate the biological consequences of antigenic variation.

Antigenic shift during natural infection

Antigenic variation which was first detected in laboratory cultures also occurs following application of an *in vivo* selection pressure generated by growth of gonococci in plastic chambers subcutaneously implanted into guinea-pigs. Chambers were infected with a mixture of variants of strain P9 and in each case new variants arose which expressed pili (Lambden *et al.*, 1981) and/or P.II (McBride *et al.*, 1981) which were not present in the original inoculum. Detection of antigenic variation during natural infection was less easily established because of the ethical need for prompt antibiotic therapy. James and Swanson (1978) reported differences in the ratio of O^+ to O^- variants associated with the stage of the menstrual cycle at which specimens were isolated. Novotny and Cownley (1980) reported differences seen by immune electron microscopy on pili in isolates from different sites in some patients and Tramont *et al.* (1979) observed a change in molecular weight and serological specificity of pili on re-infection of a patient from apparently the same primary source.

In order to investigate the occurrence of antigenic shift during the course of the natural infection, we have examined isolates taken from different sites in groups of sexual partners. Gonococci were cultured from the urethra of male patients and from the urethra and cervix of the female partners, and surface antigenic preparations were examined by SDS–PAGE (Zak *et al.*, 1984). With each group of patients considerable variation could be seen in the expression of both P.II and pili. Thus, for example, while three isolates from one group expressed a P.II of molecular weight 28 K, the isolate from the male urethra had an additional P.II of 32 K and that from the female cervix one of 30 K. In addition, each variant produced pili of distinct molecular weight, either 18.7 K, 18.5 K or 18.0 K. Indeed the molecular weight of pili differed between the isolates from the cervix and urethra of all female patients examined (Duckworth *et al.*, 1983). Thus antigenic variation of both pili and P.II appears to occur commonly during the course of the natural infection. This view is strengthened by observations on series of isolates obtained during an outbreak caused by a penicillin-resistant strain of gonococcus (Schwalbe *et al.*, 1985). Variants expressed zero, one, two or three P.II species from a total of at least seven and gonococci recovered from the same patient at different times had different P.II profiles. Corresponding structural differences were also seen in pili expressed by sequential isolates from an individual in the same outbreak (Hagblom *et al.*, 1985). The occurrence of extensive variation in both of the two reported studies suggests that antigenic shift in pilus and P.II expression generally occurs during the course of the natural infection and must play an important role in pathogenesis of gonococcal disease.

Antigenic variation in other surface proteins

A variety of reports suggest that variations may occur in surface antigens other than pili and P.II. Several laboratories have reported microscopical evidence that under certain growth conditions gonococci may produce a capsule (Hendley *et al.*, 1977; Richardson and Sadoff, 1977; James and Swanson, 1977; Demarco de Hormaeche *et al.*, 1978) but these observations have been difficult to reproduce and no capsular material has yet been purified. A subsequent study suggests that the morphological appearance of a capsule might be related to changes in lipopolysaccharide (LPS) structure (Demarco de Hormaeche *et al.*, 1983). Changes in LPS structure, as revealed by pyocin sensitivity, may also be associated with the increased resistance to the bactericidal effect of human sera which can be induced by treatment of gonococci with a low molecular weight fraction of guinea pig serum (Winstanley *et al.*, 1984). Increased resistance to phagocytic killing of an *in vivo* selected variant has been associated with production of protein(s) with a molecular weight of about 20 K (Parsons *et al.*, 1985). One report also suggests that a switch in P.I expression to a different immunochemical type may have been observed in spontaneous mutants of one strain (L.F.Guymon quoted by Cannon and Sparling, 1984). The precise significance of these other potential variations in surface antigen expression must await more detailed molecular studies and the remainder of this chapter will be devoted to the structural, immunochemical and biological investigation of antigenic variation in pilus and P.II expression.

Structural and immunochemical studies of variant proteins

Protein II

Peptide mapping studies with P.II species from different strains have shown that the different molecular species share structural homology (Swanson, 1980). Similarly when different P.II species from strain P9 were labelled with ^{125}I in the presence of SDS to unfold the protein, tryptic peptide maps showed considerable similarity with only a few peptides unique to any one protein (Heckels, 1981). In contrast, autoradiographs of peptide maps of P.IIs from variants which had been specifically surface labelled with ^{125}I showed fewer spots and considerable differences, with the major labelled peptides unique to an individual P.II type. Thus the different P.II species produced by a strain apparently form a family of structurally related proteins, with a common region located in the membrane and a variable surface-exposed region (Heckels, 1981). Similar studies with five distinct P.IIs from strain JS3 produced similar conclusions but also identified some common surface-exposed peptides (Judd, 1985).

The variations in surface structure of P.II are associated with alterations in antigenic specificity. Polyclonal antiserum directed against P.II can be produced by immunization of rabbits with outer membranes from a P.II$^+$ variant followed by absorption of the resulting sera with membranes from a P.II$^-$ variant. Quantitative immunoassay with such antisera raised against P.IIa from strain P9 showed less than 5% cross-reactivity with different P.II species produced by other variants of the same strain (Diaz and Heckels, 1982). Swanson and Barrera (1983a) obtained similar data using radioimmunoprecipitation and Western blotting with antisera raised against whole gonococci, and also found, in contrast, that immunization with isolated and denatured P.II produced cross-reacting antibodies. Thus, despite conserved structural domains, native P.II molecules on the gonococcal surface adopt conformations such that the variable domains become immunodominant and hence the proteins are antigenically distinct.

Pili

Pili can be obtained from gonococci by repeated cycles of disaggregation and precipitation (Brinton *et al.*, 1978) and are found to be composed of a repeating array of pilin subunits which have molecular weights in the range approximately 17−22 K. Variant α, β, γ and δ pili from strain P9 with subunit molecular weights of 19.5, 20.5, 21 and 18.5 K have been purified and shown by amino acid analysis to have considerable similarity. Peptide maps of tryptic/chymotryptic digests of the ^{125}I-labelled pili showed considerable structural homology, with several major peptides in common but with some unique to a particular pilus type (Lambden, 1982). Despite common determinants, rabbit antisera raised against each pilus type showed less than 10% cross-reactivity with the heterologous types (Virji *et al.*, 1982). Thus, like P.II, structural variations occur in immunodominant domains which generate antigenically distinct molecules.

Structural studies by Schoolnik and colleagues with pili from strains R10 and MS-11 have provided insight into the structural basis of inter-strain antigenic diversity (Schoolnik *et al.*, 1983, 1984). Pili contain only two methionine residues and so can be cleaved by cyanogen bromide (CNBr) to produce three peptides, CNBr-1 comprising the first seven residues from the N terminus, CNBr-2 (residues 8−92) and CNBr-3 (residues 93− 159) (*Figure 4*). Sequence analysis of CNBr-2 obtained from pili of the two strains

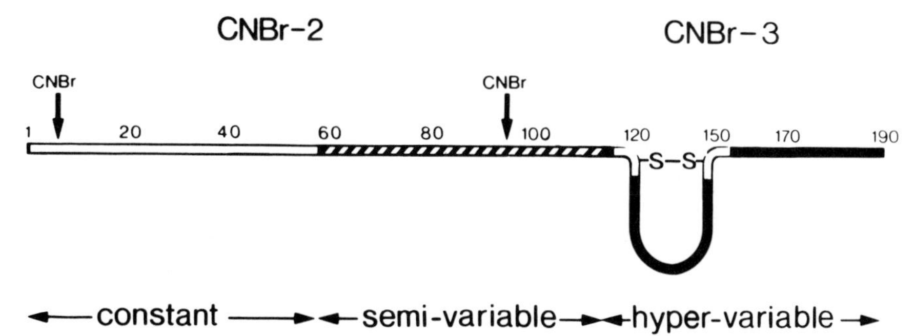

Fig. 4. Schematic diagram of pilin structure showing constant ▭, semivariable ▨ and hypervariable ■ regions. Arrows show methionine residues which are the CNBr cleavage points. Based on pilin amino acid sequence analysis (Schoolnik *et al.*, 1984) and primer extension sequencing of mRNA (Hagblom *et al.*, 1985) produced by variants of strain MS-11.

Table 1. Cross-reaction of monoclonal antibodies with variant pili from three strains.

Antibody	Strain P9				Group 1			Group 2		
	α	β	γ	δ	M	Fc	Fu	M	Fc	Fu
SM1	+	+	+	+	+	+	+	+	+	+
SM2	+	+	+	+	+	+	+	+	+	+
SM3	+	−	−	−	−	−	−	−	−	−
SM4	+	−	−	−	−	−	−	−	−	−
SM5	+	−	−	−	−	−	−	−	−	−
SM6	−	−	+	+	−	−	−	−	−	−
SM7	−	−	+	+	−	−	−	−	−	−
SM8	−	−	+	−	−	−	−	−	−	−
SM10	+	+	−	+	+	−	−	−	−	−
SM11	+	−	−	−	−	−	−	+	−	−
SM13	+	−	+	−	nd	nd	nd	nd	nd	nd

Pili were purified from variants of strain P9 and from two groups of consorts and used in enzyme-linked immunosorbent assay (ELISA) with a panel of monoclonal antibodies raised against variants of strain P9. M, isolate from urethra of male patient; F, isolates from their female partners (u = urethra, c = cervix). (From Virji and Heckels, 1983.)

showed homology through to residue 59. In contrast, the CNBr-3 fragments showed significant differences between the strains, particularly in the region encompassed by the disulphide loop. The authors proposed a model in which pili contained a constant region at the N terminus and a variable region at the C terminus. It was also found that antibodies raised against intact pili are predominantly directed against epitopes within CNBr-3 and that the conserved region of CNBr-2 is immunorecessive (Rothbard *et al.*, 1984). Hence structural differences in the variable region result in antigenically distinct pili despite the occurrence of considerable regions of structural homology.

We have studied the immunochemical differences between variant pili produced by a single strain using monoclonal antibodies. A panel of monoclonal antibodies was obtained with activity against the variant α, β, γ and δ pili from strain P9 (Virji *et al.*, 1983). Two types of activity were observed (*Table 1*). Only two antibodies (SM1 and SM2), from over 100 screened, reacted with each of the pilus types from P9 and also with pili from all other strains tested. In contrast, the remaining antibodies showed specificity for only one or two of the P9 variant pili and only occasional reactivity with pili from any other strains (Virji and Heckels, 1983). The distribution of the epitopes was investigated by competitive radioimmunoassay. With purified α pili the two cross-reacting antibodies were not inhibited by any of the type-specific antibodies or vice versa, while some pairs of type-specific antibodies showed significant inhibition and others did not. By analogy with the model of Schoolnik *et al.* (1983) for inter-strain variation we proposed that variant pili contain two antigenic regions, a common region present on each pilus type and an antigenically distinct region responsible for type specificity. Both sites contain more than one epitope and pilus type specificity is determined by a particular combination from a large number of possible epitopes which are shared both within and between strains (Virji and Heckels, 1983) (*Figure 4*). Subsequent studies have revealed that although antibodies SM1 and SM2 show mutual inhibition they recognize different CNBr fragments (Virji and Heckels, 1985a). Antibody SM1 reacts with the common CNBr-2 but SM2 recognizes a conserved determinant with the variable CNBr-3. Thus conserved domains occur even within the variable region responsible for antigenic specificity.

These conclusions are in accord with sequencing studies carried out by So and colleagues on mRNA from pili obtained from variants of strains MS-11 (Meyer *et al.*, 1984; Hagblom *et al.*, 1985). They suggested that pili can be considered to comprise three regions, a constant region comprising the first 53 amino acids, a semivariable (regions 54–114) and a hypervariable region at the carboxy terminus (*Figure 4*). Structural variations in the semivariable region arise from amino acid substitutions, but in the hypervariable region insertions and deletions of up to four amino acids occur hence generating antigenic diversity. Even within the hypervariable region two conserved sequences occur centred around the two cysteine residues at positions 121 and 151. Genetic studies suggest that antigenic diversity is generated as a result of assembly at an expression site of separated silent gene segments (M.So, personal communication; see Saunders, Chapter 5).

Effect of antigenic variation on pathogenic properties

The occurrence of structural and antigenic differences in important surface determinants might be expected to exert a significant influence on gonococcal pathogenesis particularly in interactions with host cells. *In vitro* model systems have been used to investigate two such interactions, namely adhesion to epithelial cells and interaction with phagocytes.

Adhesion to epithelial cells

The initial association between piliation and virulence (see above) has led to development of several model systems to assess pilus binding to epithelial cells. One convenient system utilizes human buccal epithelial cells, since it has been reported that they contain a similar pilus receptor density to the epithelial cells of the genital tract (Pearce and Buchanan, 1978). When the binding of purified α and β pili from strain P9 were compared in this system, striking differences were seen (Lambden *et al.*, 1980). Binding of α pili was pH dependent with a maximum at pH 6.5 whereas attachment of β pili was considerably lower and showed no pH optimum. Further differences in adhesive properties were revealed using buccal cells after incubation with glycosidases, a treatment which reduced binding of α pili to the level of β pili. Similar results were obtained with intact piliated gonococci (Trust *et al.*, 1980) and it was suggested that α pili recognize an oligosaccharide receptor present on the surface of buccal cells and that β pili lack the appropriate ligand. An alternative method uses Chang conjunctival epithelial cells, growing in tissue culture, to mimic the ability of gonococci to adhere to and destroy epithelial cells of the genital tract (Virji and Everson, 1981). In this system, in contrast to the buccal cell model, β and γ piliated variants showed much greater adhesion and hence virulence than did α and δ piliated variants (Virji *et al.*, 1982). Hence variant pili may show altered specificity for different cell types in *in vitro* model systems suggesting that *in vivo* variation might endow gonococci with the ability to colonize different cell types.

Variations in expression of P.II also influence gonococcal adhesion in model systems. Comparison of variants of strain P9 showed that all P.II$^+$ variants exhibit increased attachment to buccal epithelial cells compared with a Pil^- $P.II^-$ control (Lambden *et al.*, 1979). Similarly increased adhesion of $P.II^+$ variants has been demonstrated using human tissue culture cell lines including Chang (Virji and Everson, 1981), HeLa (Sugasawara *et al.*, 1983) and Flow 2000 cells (James *et al.*, 1980). The contribution of P.II to adhesion has been investigated using purified outer membranes in the buccal epithelial cell model (Heckels, 1982). Results showed that membranes containing P.IIa had a 5-fold greater avidity of attachment than membranes from the $P.II^-$ variant. This advantage was destroyed by treating either the cells with glycosidases or the membranes with trypsin to degrade P.II. These observations confirm the contribution of P.II in adhesion to buccal cells and suggest that host cell surface carbohydrates are involved in the adhesive process. In contrast, $P.II^+$ variants are reported to have decreased association with human fallopian tube and uterine explants (James *et al.*, 1980; Draper *et al.*, 1980a).

Specificity of adhesion is shown not only between $P.II^+$ and $P.II^-$ variants but also between different $P.II^+$ variants of the same strain. The different molecular forms of P.II expressed by strain P9 were not equal in their effect in the buccal cell model, with the variant expressing P.IIa giving the greatest adhesion and P.IIb the least (Lambden *et al.*, 1979). With Chang conjuctiva cells the position was reversed, the variant expressing P.IIb showed greatest adhesion and hence virulence for the cell line (Virji and Everson, 1981). Thus, as with pili, structural variations in P.II may influence the ability of gonococci to colonize different anatomic sites.

Interaction with polymorphonuclear leukocytes

The interaction of gonococci with PMN has important consequences for the eventual outcome of an infection. Pili have been implicated in resistance to phagocytosis by several studies (Thongthai and Sawyer, 1973; Gibbs and Roberts, 1975; Dilworth *et al.*, 1975; Densen and Mandell, 1978) while others have suggested that small numbers of Pil^+ variants enter phagocytic cells but then resist intracellular killing (Thomas *et al.*, 1973; Witt *et al.*, 1976). Swanson and co-workers, on the other hand, suggested that pili had only a minor effect on gonococal–PMN interactions (Swanson *et al.*, 1974) and that an outer membrane component termed leukocyte association (LA) factor played the predominant role (Swanson *et al.*, 1975). Variants of strain MS-11 which exhibited LA^+ behaviour had increased amounts of surface proteins of 29 K and 28 K which were termed leukocyte association proteins (Swanson and King, 1978). These proteins exhibited biochemical properties similar to those later recognized as characteristic of P.II species (Swanson and Heckels, 1980). Subsequently two studies showed that Pil^- $P.II^+$ variants showed increased leukocyte interactions compared with their $Pil^-P.II^-$ counterparts (Lambden *et al.*, 1979; Rest *et al.*, 1982).

Questions which remained to be answered were whether LA proteins formed a special subset of P.II, or if all P.IIs possessed the property to some degree, and whether pilus variation might have an additional effect. We therefore utilized a panel of variants of strain P9 with defined differences in pili and/or P.II in a chemiluminescence (CL) assay to determine initial interactions and a phagocytic killing assay to determine the ultimate fate of the organisms (Virji and Heckels, 1986). In this study LA^+ behaviour was synonymous with possession of P.II since all $P.II^+$ variants showed increased interaction and all were readily killed by PMN, in contrast to P.II$^-$ variants which were resistant to phagocytic killing. The P.II species were not, however, equal in their effect and the shape of the CL response curve differed between variants. When pairs of variants expressing the same P.II but either Pil$^+$ or Pil$^-$ were compared, in each case the CL response was determined by the particular molecular species of P.II present and pili had an insignificant effect. Moreover, pili did not inhibit either uptake or intracellular killing, confirming the predominant role of P.II in PMN interactions and suggesting that pili play a small role in resistance to phagocytosis by PMN. The importance of P.II in the interaction was also revealed by the ability of monoclonal $F(ab')_2$ directed against P.IIb to inhibit CL of the variant containing the protein (Virji and Heckels, 1986). It is interesting to note that the LA$^+$ characteristic does not influence gonococcal−macrophage interactions (Blake and Swanson, 1975) and that P.II has little effect on overall surface hydrophobicity (Lambden *et al.*, 1979; Magnusson *et al.*, 1979). Thus it would appear that the increased gonococcal−PMN interactions seen with P.II$^+$ variants may result from specific interactions rather than a change in the gross physicochemical properties of the bacterial surface.

The effect of antibodies directed against variant proteins

The importance of variable proteins in pathogenesis has also been revealed by use of antibodies directed against pili and P.II. Early model studies utilized antisera raised by immunization with purified or partially purified pilus preparations. The antisera obtained reduced adhesion of piliated gonococci to human buccal epithelial cells (Tramont, 1976) and of purified pili to both buccal cells (Pearce and Buchanan, 1978) and erythrocytes (Buchanan and Pearce, 1976). Antibodies to pili also opsonized gonococci for phagocytosis by macrophages (Jones *et al.*, 1980) and PMN (Punsalang and Sawyer, 1973). However, in most of the above studies optimal activity was only seen when the test and immunizing strains were the same and the additional problems posed by intrastrain variation were not investigated. In one study, protection of Chang conjunctiva cells against variants of strain P9 was obtained with the homologous variant while antibodies to the other variant pili produced only limited protection or none at all (Virji *et al.*, 1982).

One complication of such protection studies with animal sera is the possible presence in polyclonal sera of low levels of protective antibodies directed against other components. The use of monoclonal antibodies avoids this possibility and permits investigation of the specific effect of antibodies directed against particular pilus epitopes. Monoclonal antibodies raised against variant pili from strain P9 have been used in several model systems. The binding of ^{125}I-labelled α pili was inhibited by three type-specific but not two cross-reacting antibodies. Four type-specific antibodies inhibited binding of γ pili while the two cross-reacting antibodies were without effect. The virulence of variants P9-2 (α piliated) and P9-35 (γ piliated) for Chang cells was similarly reduced only in the presence of the relevant type-specific antibodies (Virji and Heckels, 1984). Type-specific antibodies were also opsonic and promoted phagocytic killing of piliated variants, but cross-reacting antibodies again showed no protective effect (Virji and Heckels, 1985).

Because of problems associated with obtaining a pure but 'native' protein, similar studies with P.II have become possible only with the advent of monoclonal antibodies. Type-specific antibodies to P.II inhibit binding of the corresponding variant to HeLa cells (Sugasawara *et al.*, 1983), protect in the Chang cell model, are opsonic and are bactericidal (Heckels and Virji, 1986). No information is yet available about the effect of antibodies directed against conserved epitopes of P.II.

The role of antigenic variation in pathogenesis

Genetic studies of pilus expression have revealed the existence of complex mechanisms generating antigenic diversity by recombination of separated silent gene segments (reviewed by Saunders, Chapter 5). The fact that a substantial proportion of the gonococcal genome is devoted to antigenic variation suggests that it must endow the gonococcus with an important survival advantage, presumably allowing adaptation to a changing external environment. During the natural infection, gonococci may colonize and invade a variety of mucosal surfaces at different anatomic locations and there encounter different physiological conditions as well as specific host defence mechanisms. The precise nature of the driving force inducing antigenic variation is unclear and it may be naive to expect that a single influence could predominate. Indeed, it is likely that the antigens expressed by a gonococcus at any stage of the infection result from a series of complex interactions between gonococci and their external environment.

One important factor must be the host immune system. Antibodies to both pili and P.II are protective in model systems, but antigenic shift would enable gonococci to evade the immune response. Evidence for antibody-mediated variation has come from a study with serum from groups of consorts (Zak *et al.*, 1984). Antibody levels were measured against each of the isolates obtained from within a patient's own contact group. Serum from one woman contained antibodies reacting with a 31-K P.II but not a 29.5-K P.II present in her urethral isolate or with the P.II present in the isolate from her male partner, confirming the specificity of the human immune response directed against P.II. The isolate from a second female contact of the male expressed a 29-K P.II. Serum from the first female contained antibodies against this protein, despite its absence from either of her own isolates, suggesting previous exposure to the antigen. Thus a model for antigenic shift is that in the initial infection the gonococcus produces one of several possible P.II variants, inducing an immune response which subsequently results in the elimination of variants expressing that protein. The inherently high rate of variation in P.II expression (Mayer, 1982) would always ensure the presence of a significant minority population, with a different P.II profile, which would then grow to establish the new majority population. The cycle could presumably be repeated several times and similar selection pressures would operate concurrently on pilus expression.

The altered specificity of adhesion shown by different molecular species of both pili and P.II in *in vitro* assay systems (see above) suggests that structural variations in either may confer tissue tropisms, allowing colonization of the different anatomic sites. No evidence is yet available to confirm this hypothesis and in two studies of variation during natural infection, no correlation could be made between site of isolation and molecular species of P.II or pili expressed (Zak *et al.*, 1984; Schwalbe *et al.*, 1985). However, the increased prevalence of P.II$^-$ variants isolated from cases of salpingitis (Draper *et al.*, 1980b) combined with their increased adhesion to fallopian tube mucosa *in vitro*

(Draper *et al.*, 1980a), suggests that loss of P.II expression may provide a selective advantage in colonization of this and perhaps other sites.

Although the majority of fresh clinical isolates possess P.II (Zak *et al.*, 1984) which clearly impart essential properties such as ability to colonize most epithelial surfaces, the ability to switch off P.II expression may increase virulence under other circumstances. Possession of P.II renders gonococci susceptible to phagocytosis *in vitro*, due to increased interaction with PMN (see above) and also since antibodies to P.II are opsonic (Heckels and Virji, 1986). Transparent variants (presumably lacking P.II) show much greater toxicity than P.II$^+$ variants for 11-day-old chick embryos (Salit and Gotschlich, 1978). When embryos were injected with P.II$^+$ variants they developed a bacteraemia caused by P.II$^-$ variants. Since chick embryos lack bactericidal activity but have an effective phagocytic system (Board and Fuller, 1974), the survival of P.II$^-$ variants may well result from their increased resistance to phagocytosis. Certainly the majority of survivors from *in vitro* phagocytosis experiments with human PMN lack P.II (Virji and Heckels, 1986), suggesting that any P.II$^-$ variants arising during the natural infection may escape phagocytosis and ultimately re-colonize other niches. An increased proportion of O$^-$ variants isolated from women near to the time of menstruation has been associated with the increased sensitivity of P.II$^+$ variants to proteolytic enzymes which attain maximum levels at this time (James and Swanson, 1978). An additional factor may also be the increased sensitivity of P.II$^+$ variants to progesterone which is at a maximum during the luteal phase of the cycle (Salit, 1982). Whatever the mechanism, the generation of P.II$^-$ variants resistant to phagocytosis by PMN may well be responsible for the observed prevalence of disseminated infections following menstruation (Holmes *et al.*, 1971) and the association of P.II$^-$ variants with such infections (O'Brien *et al.*, 1983).

Role of conserved structural domains in variant proteins

A primary goal of a gonococcal vaccine would be to prevent initial colonization of the mucosal surfaces of the genital tract. The only surface components so far implicated in the adhesive process are pili and P.II. Clearly the phenomen of antigenic shift combined with the immunodominance of their variable determinants makes them appear unattractive vaccine candidates, even though antibodies to either antigen inhibit adhesion of the homologous variants. Nevertheless, common determinants do elicit low but significant levels of cross-reacting antibodies on immunization, and common domains may contain important functionally conserved regions. One approach is therefore to devise strategies to increase the antibody response to the conserved domains in the hope that such antibodies would exert a protective effect. Little information is available on the biological role of the conserved regions of P.II but studies on pili have produced interesting information.

Studies with peptides CNBr-2 and CNBr-3 (*Figure 4*) derived by cleavage of pili have shown that the common fragment CNBr-2 has the ability to bind to erythrocytes and tissue culture cells, suggesting that it might contain a functionally conserved domain involved in cell recognition (Schoolnik *et al.*, 1983; Gubish *et al.*, 1982). We have used monoclonal antibody SM1 (*Table 1*) to examine the potential protective effect of antibodies reacting with the CNBr-2 (Virji and Heckels, 1984). The results were disappointing in that, unlike antibodies directed against variable determinants in CNBr-3, it failed to inhibit adherence of pili to buccal cells, did not reduce gonococcal virulence for Chang cells and was not opsonic. Further studies have revealed that antibody SM1 recognizes an epitope between amino acid residues 48 and 60 of the pilin molecule (Heckels and Virji, 1986). The low levels of cross-reacting antibodies obtained on immunization with intact pili also react with this epitope (Rothbard *et al.*, 1984), and antibodies raised by immunization with synthetic 48–60 also fail to inhibit pilus binding to an endocervical cell line (Rothbard *et al.*, 1985). Thus gonococcal pili appear to have evolved so that the main immune response to pili is directed against variable determinants and even the low levels of cross-reacting antibodies are directed against non-protective epitopes. This is an effective strategy for the gonococcus but clearly poses considerable problems for the development of a pilus vaccine.

Nevertheless, considerable progress has been made by an alternative approach developed by Schoolnik and colleagues using synthetic peptides. Immunization of rabbits with the CNBr-2 fragment produces antibodies directed primarily against peptides 41–50 and 69–84, rather than the ineffective 48–60 (Rothbard *et al.*, 1984), and these antibodies inhibit pilus binding to erythrocytes (Schoolnik *et al.*, 1983). Synthetic peptides 41–50 and 69–84 have been coupled to a carrier protein and used to immunize rabbits, and the antisera obtained inhibited adhesion of piliated gonococci to a human endometrial carcinoma cell line and were effective against a heterologous strain (Rothbard *et al.*, 1985). Thus although pilus adhesion to epithelial cells may involve both conserved and variable determinants, cross-reacting antibodies which block adhesion *in vitro* can be raised using a common synthetic immunogen. Clearly these studies show considerable promise in the search for suitable components for a gonococcal vaccine.

Conclusions

Molecular and biological studies of gonococcal pathogenesis have provided a fascinating insight into the life-style of an extremely effective pathogen. The face which the gonococcus presents to the outside world has proved to be enormously variable. The variations in pili and P.II are associated with alterations in important virulence properties, suggesting that antigenic shift plays a crucial role in the success of gonococci, enabling them to adapt to an ever-changing host environment. Colonization of a variety of mucosal surfaces and evasion of host defences are important characteristics of gonococcal infections which are influenced by surface antigen expression. Antigenic variation is clearly a major obstacle to the development of an effective vaccine but the identification and isolation of functionally conserved domains provides an attractive strategy for current vaccine research.

Acknowledgements

I am grateful to Dr M.So for providing unpublished observations. Work carried out in Southampton was supported by the Medical Research Council and the World Health Organisation and has been carried out in collaboration with Drs M.Virji and P.R. Lambden.

References

Black,W.J., Schwalbe,R.S., Nachamkin,I. and Cannon,J.G. (1984) Characterisation of *Neisseria gonorrhoeae* protein II phase variation by use of monoclonal antibodies. *Infect. Immun.*, **45**, 453–457.

Blake,M.S. and Gotschlich,E.C. (1983) Gonococcal membrane proteins: speculation on their role in pathogenesis. *Prog. Allergy*, **33**, 298–313.

Blake,M. and Swanson,J. (1975) Studies on gonococcus infection. IX. *In vitro* decreased association of piliated gonococci with mouse peritoneal macrophages. *Infect. Immun.*, **11**, 1402–1404.

Board,R.G. and Fuller,R. (1974) Non-specific antimicrobial defences of the avian egg, embryo and neonate. *Biol. Rev.*, **49**, 15–49.

Brinton,C.C., Bryan,J., Dillon,J.-A., Guerina,N., Jacobson,L.J., Labik,A., Lee,S., Levine,A., Lim,S., McMichael,J., Polen,S., Rogers,K., To,A.C.-C. and To,S.C.-M. (1978) Uses of pili in gonorrhoea control: role of bacterial pili in disease, purification and properties of gonococcal pili and progress in the development of a gonococcal pilus vaccine for gonorrhea. In *Immunobiology of Neisseria gonorrhoeae*. Brooks,G.F., Gotschlich,E.C., Holmes,K.K., Sawyer,W.D. and Young,F.E. (eds), American Society for Microbiology, Washington, DC, pp. 155–178.

Buchanan,T.M. and Pearce,W.C. (1976) Pili as a mediator of the attachment of gonococci to human erythrocytes. *Infect. Immun.*, **13**, 1483–1489.

Cannon,J.G. and Sparling,P.F. (1984) The genetics of the gonococcus. *Annu. Rev. Microbiol.*, **38**, 111–133.

Demarco de Hormaeche,R., Thornley,M.J. and Glauert,A.M. (1978) Demonstration by light and electron microscopy of capsules on gonococci recently grown *in vivo*. *J. Gen. Microbiol.*, **106**, 81–91.

Demarco de Hormaeche,R., Thornley,M.J. and Holmes,A. (1983) Surface antigens of gonococci: correlation with virulence and serum resistance. *J. Gen. Microbiol.*, **129**, 1559–1567.

Densen,P. and Mandell,G.L. (1978) Gonococcal interactions with polymorphonuclear neutrophils: importance of the phagosome for bactericidal activity. *J. Clin. Invest.*, **62**, 1161–1171.

Diaz,J.-L. and Heckels,J.E. (1982) Antigenic variation of outer membrane protein II in colonial variants of *Neisseria gonorrhoeae* P9. *J. Gen. Microbiol.*, **128**, 585–591.

Dilworth,J.A., Hendley,J.O. and Mandell,G.L. (1975) Attachment and ingestion of gonococci by human neutrophils. *Infect. Immun.*, **11**, 512–516.

Douglas,J.T., Lee,M.D. and Nikaido,H. (1981) Protein I of *Neisseria gonorrhoeae* outer membrane is a porin. *FEMS Microbiol. Lett.*, **12**, 305–309.

Draper,D.L., Donegan,E.A., James,J.F., Sweet,R.L. and Brooks,G.F. (1980a) *In vitro* modeling of acute salpingitis caused by *Neisseria gonorrhoeae. Am. J. Obstet. Gynaecol.*, **138**, 996–1002.

Draper,D.L., James,J.F., Brooks,G.F., Jr. and Sweet,R.L. (1980b) Comparison of virulence markers of peritoneal and fallopian tube isolates with endocervical *Neisseria gonorrhoeae* isolates from women with acute salpingitis. *Infect. Immun.*, **27**, 882–888.

Duckworth,M., Jackson,D., Zak,K. and Heckels,J.E. (1983) Structural variations in pili expressed during gonococcal infection. *J. Gen. Microbiol.*, **129**, 1593–1596.

Gibbs,D.L. and Roberts,R.B. (1975) The interaction *in vitro* between human polymorphonuclear leukocytes and *Neisseria gonorrhoeae* cultivated in the chick embyro. *J. Exp. Med.*, **141**, 155–171.

Gubish,E.R., Chen,K.C.S. and Buchanan,T.M. (1982) Attachment of gonococcal pili to lectin-resistant clones of Chinese hamster ovary cells. *Infect. Immun.*, **37**, 189–194.

Hagblom,P., Segal,E., Billyard,E. and So,M. (1985) Intragenic recombination leads to pilus antigenic variation in *Neisseria gonorrhoeae. Nature*, **315**, 156–158.

Harkness,A.H. (1948) The pathology of gonorrhoea. *Br. J. Vener. Dis.*, **24**, 137–147.

Heckels,J.E. (1977) The surface properties of *Neisseria gonorrhoeae*: isolation of the major components of the outer membrane. *J. Gen. Microbiol.*, **99**, 333–341.

Heckels,J.E. (1978) The surface properties of *Neisseria gonorrhoeae*: topographical distribution of the outer membrane protein antigens. *J. Gen. Microbiol.*, **108**, 213–219.

Heckels,J.E. (1979) The outer membrane of *Neisseria gonorrhoeae*: evidence that protein I is a transmembrane protein. *FEMS Microbiol. Lett.*, **6**, 325–327.

Heckels,J.E. (1981) Structural comparison of *Neisseria gonorrhoeae* outer membrane proteins. *J. Bacteriol.*, **145**, 736–742.

Heckels,J.E. (1982) Role of surface proteins in the adhesion of *Neisseria gonorrhoeae*. In *Microbiology 1982*. Schlessinger,D. (ed.), American Society for Microbiology, Washington, DC, pp. 301–304.

Heckels,J.E. and Virji,M. (1986) Antigenic variation of gonococcal surface proteins: effect on virulence. In *Molecular Biology of Microbial Pathogenicity*. Normark,S. (ed.), Academic Press, London, in press.

Heckels,J.E., Blackett,B., Everson,J.S. and Ward,M.E. (1976) The influence of surface charge on the attachment of *Neisseria gonorrhoeae* to human cells. *J. Gen. Microbiol.*, **96**, 359–364.

Hendley,J.O., Powell,K.R., Rodewald,R., Halzgrefe,H.H. and Lyles,R. (1977) Demonstration of a capsule on *Neisseria gonorrhoeae*. *New Engl. J. Med.*, **296**, 608–611.

Holmes,K.K., Courts,G.W. and Beatty,H.N. (1971) Disseminated gonococcal infection. *Ann. Internal Med.*, **74**, 979–993.

James,J.F. and Swanson,J. (1977) The capsule of the gonococcus. *J. Exp. Med.*, **145**, 1082–1086.

James,J.F. and Swanson,J. (1978) Studies on gonococcus infections. XIII. Occurrence of color/opacity colonial variants in clinical cultures. *Infect. Immun.*, **19**, 332–340.

James,J.F., Lammel,C.J., Draper,D.L. and Brooks,G.F. (1980) Attachment of *N. gonorrhoeae* colony phenotype variants to eukaryotic cells and tissues. In *Genetics and Immunobiology of Pathogenic Neisseria.* Danielsson,D. and Normark,S. (eds), University of Umea, Sweden, pp. 213–216.

Jephcott,A.E., Reyn,A. and Birch-Andersen,A. (1971) Brief report: *Neisseria gonorrhoeae*. III. Demonstration of presumed appendages to cells from different colony types. *Acta Pathol. Microbiol. Scand. Sect. B*, **79**, 437–439.

Johnston,K.H. and Gotschlich,E.C. (1974) Isolation and characterisation of the outer membrane of *Neisseria gonorrhoeae*. *J. Bacteriol.*, **119**, 250–257.

Johnston,K.H., Holmes,K.K. and Gotschlich,E.C. (1976) The serological classification of *Neisseria gonorrhoeae*. I. Isolation of the outer membrane complex responsible for serotypic specificity. *J. Exp. Med.*, **143**, 741–758.

Jones,R.B., Newland,J.C., Olsen,D.A. and Buchanan,T.M. (1980) Immune enhanced phagocytosis of *Neisseria gonorrhoeae* by macrophages: characterisation of the major antigens to which opsonins are directed. *J. Gen. Microbiol.*, **121**, 365–372.

Judd,R.C. (1982) ^{125}I-peptide mapping of protein III isolated from four strains of *Neisseria gonorrhoeae*. *Infect. Immun.*, **37**, 622–631.

Judd,R.C. (1985) Structure and surface exposure of protein IIs of *Neisseria gonorrhoeae*. *Infect. Immun.*, **48**, 452–457.

Kellog,D.S., Peacock,W.L., Deacon,W.E., Brown,L. and Pirkle,C.I. (1963) *Neisseria gonorrhoeae*. I. virulence genetically linked to clonal variation. *J. Bacteriol.*, **85**, 1274–1279.

Lambden,P.R. (1982) Biochemical comparisons of pili from variants of *Neisseria gonorrhoeae* P9. *J. Gen. Microbiol.*, **128**, 2105–2111.

Lambden,P.R. and Heckels,J.E. (1979) Outer membrane protein composition and colonial morphology of *Neisseria gonorrhoeae* strain P9. *FEMS Microbiol. Lett.*, **5**, 263–265.

Lambden,P.R., Heckels,J.E., James,L.T. and Watt,P.J. (1979) Variations in surface protein composition associated with virulence properties in opacity types of *Neisseria gonorrhoeae*. *J. Gen. Microbiol.*, **114**, 305–312.

Lambden,P.R., Robertson,J.N. and Watt,P.J. (1980) Biological properties of two distinct pilus types produced by isogenic variants of *Neisseria gonorrhoeae* P9. *J. Bacteriol.*, **141**, 393–396.

Lambden,P.R., Heckels,J.E., McBride,H. and Watt,P.J. (1981) The identification and isolation of novel pilus types produced by variants of *N. gonorrhoeae* P9 following selection *in vivo*. *FEMS Microbiol. Lett.*, **10**, 339–341.

Lynch,E.C., Blake,M.S., Gotschlich,E.C. and Mauro,A. (1984) Studies of porins: spontaneously transferred from whole cells and reconstituted from purified proteins of *Neisseria gonorrhoeae* and *Neisseria meningitidis*. *Biophys. J.*, **45**, 104–107.

Magnusson,K.E., Kihlstrom,E., Norlander,L., Norquist,A., Davies,J. and Normark,S. (1979) Effect of colony type and pH on surface charge and hydrophobicity of *Neisseria gonorrhoeae*. *Infect. Immun.*, **26**, 397–401.

Mardh,P.A. and Westrom,L. (1976) Adherence of bacteria to vaginal epithelial cells. *Infect. Immun.*, **13**, 661–666.

Mayer,L.W. (1982) Rates of *in vitro* changes in gonococcal colony opacity phenotypes. *Infect. Immun.*, **37**, 481–485.

McBride,H.M., Lambden,P.R., Heckels,J.E. and Watt,P.J. (1981) The role of outer membrane proteins in the survival of *Neisseria gonorrhoeae* P9 within guinea pig subcutaneous chambers. *J. Gen. Microbiol.*, **126**, 63–67.

McDade,R.L. and Johnson,K.H. (1980) Characterisation of serologically dominant outer membrane proteins of *Neisseria gonorrhoeae*. *J. Bacteriol.*, **141**, 1183–1191.

Meyer,T.F., Billyard,E., Haas,R., Strozbach,S. and So,M. (1984) Pilus genes of *Neisseria gonorrhoeae*: chromosomal organisation and DNA sequence. *Proc. Natl. Acad. Sci. USA*, **81**, 6110–6114.

Newhall,W.J., Sawyer,W.D. and Haak,R.A. (1980) Cross-linking analysis of the outer membrane proteins of *Neisseria gonorrhoeae. Infect. Immun.*, **28**, 785–791.

Novotny,P. and Cownley,K. (1980) Investigation of serological cross-reactivity between gonococcal pili. In *Genetics and Immunobiology of Pathogenic Neisseria.* Normark,S. and Danielsson,D. (eds), University of Umea, Sweden, pp. 91–94.

O'Brien,J.P., Goldenberg,D.L. and Rice,P.A. (1983) Disseminated gonococcal infection: a prospective analysis of 49 patients and a review of pathophysiology and immune mechanisms. *Medicine,* **62**, 395–406.

Parsons,N.J., Kwaasi,A.A.A., Patel,P.V., Martin,P.M.V. and Smith,H. (1985) Association of resistance of *Neisseria gonorrhoeae* to killing by human phagocytes with outer membrane proteins of about 20 kilodaltons. *J. Gen. Microbiol.*, **131**, 601–610.

Pearce,W.A. and Buchanan,T.M. (1978) Attachment role of gonococcal pili. Optimum conditions and quantitation of adherence of isolated pili to human cells *in vitro. J. Clin. Invest.*, **61**, 931–943.

Perera,V.Y., Penn,C.W. and Smith,H. (1982) Variant pili of autoagglutinating *Neisseria gonorrhoeae. FEMS Microbiol. Lett.*, **13**, 313–316.

Punsalang,A.P. and Sawyer,W.D. (1973) Role of pili in the virulence of *Neisseria gonorrhoeae. Infect. Immun.*, **8**, 255–263.

Rest,R.F., Fischer,S.H., Ingham,Z.Z. and Jones,J.F. (1982) Interactions of *Neisseria gonorrhoeae* with human neutrophils: effects of serum and gonococcal opacity on phagocyte killing and chemiluminescence. *Infect. Immun.*, **36**, 737–744.

Richardson,W.P. and Sadoff,J.C. (1977) Production of a capsule by *Neisseria gonorrhoeae. Infect. Immun.*, **15**, 663–664.

Rothbard,J.B., Fernandez,R. and Schoolnik,G.K. (1984) Strain-specific and common epitopes of gonococcal pili. *J. Exp. Med.*, **160**, 208–221.

Rothbard,J.B., Fernandez,R., Wang,L., Teng,N.N.H. and Schoolnik,G.K. (1985) Antibodies to peptides corresponding to a conserved sequence of gonococcal pilins block bacterial adhesion. *Proc. Natl. Acad. Sci. USA,* **82**, 915–919.

Salit,I.E. (1982) The differential susceptibility of gonococcal colony opacity variants to sex hormones. *Can. J. Microbiol.*, **28**, 301–306.

Salit,I.E. and Gotschlich,E.C. (1978) Gonococcal color and opacity variants: virulence for chicken embyros. *Infect. Immun.*, **22**, 359–364.

Salit,I.E., Blake,M. and Gotschlich,E.C. (1980) Intra-strain heterogeneity of gonococcal pili is related to opacity colony variance. *J. Exp. Med.*, **151**, 716–725.

Sandstrom,E.G., Chen,K.C.S. and Buchanan,T.M. (1982) Serology of *Neisseria gonorrhoeae*: coagglutination serogroups W1 and WII/III correspond to different outer membrane protein molecules. *Infect. Immun.*, **38**, 462–470.

Schoolnik,G.K., Tai,J.Y. and Gotschlich,E.C. (1983) A pilus peptide for the prevention of gonorrhea. *Prog. Allergy,* **33**, 314–331.

Schoolnik,G.K., Fernandez,R., Tai,J.Y., Rothbard,J. and Gotschlich,E.C. (1984) Gonococcal pili: primary structure and receptor binding domain. *J. Exp. Med.*, **159**, 1351–1370.

Schwalbe,R.S., Sparling,P.F. and Cannon,J.G. (1985) Variation of *Neisseria gonorrhoeae* protein II among isolates from an outbreak caused by a single gonococcal strain. *Infect. Immun.*, **49**, 250–252.

Sugasawara,R.J., Cannon,J.G., Black,W.J., Nachamkin,I., Sweck,R.L. and Brooks,G.F. (1983) Inhibition of *Neisseria gonorrhoeae* attachment to HeLa cells with monoclonal antibody directed against a protein II. *Infect. Immun.*, **42**, 980–985.

Swanson,J. (1973) Studies on gonococcus infection. IV. Pili: their role in attachment of gonococci to tissue culture cells. *J. Exp. Med.*, **137**, 571–589.

Swanson,J. (1978) Studies on gonococcus infection. XIV. Cell wall protein differences among color/opacity colony variants of *Neisseria gonorrhoeae. Infect. Immun.*, **21**, 292–302.

Swanson,J. (1980) ^{125}I-labeled peptide mapping of some heat-modifiable proteins of the gonococcal outer membrane. *Infect. Immun.*, **28**, 54–64.

Swanson,J. (1982) Colony opacity and protein II compositions of gonococci. *Infect. Immun.*, **37**, 359–368.

Swanson,J. and Barrera,O. (1983a) Immunochemical characteristics of protein II assessed by immunoprecipitation, immunoblotting and coagglutination. *J. Exp. Med.*, **157**, 1405–1420.

Swanson,J. and Barrera,O. (1983b) Gonococcal pilus subunit size heterogeneity correlates with transitions in colony piliation phenotype, not with changes in colony opacity. *J. Exp. Med.*, **158**, 1459–1472.

Swanson,J. and Heckels,J.E. (1980) Proposal: nomenclature of gonococcal outer membrane proteins. In *Genetics and Immunobiology of Pathogenic Neisseria.* Danielsson,D. and Normark,S. (eds), University of Umea, Sweden, pp. xxi–xxiii.

Swanson,J. and King,G. (1978) *Neisseria gonorrhoeae* — granulocyte interactions. In *Immunobiology of Neisseria gonorrhoeae*. Brooks,G.F., Gotschlich,E.C., Holmes,K.K., Sawyer,W.D. and Young,F.E. (eds), American Society for Microbiology, Washington, DC, pp. 221–226.

Swanson,J., Kraus,S.J. and Gotschlich,E.C. (1971) Studies on gonococcus infection. I. Pili and zones of adhesion: their relation to gonococcal growth patterns. *J. Exp. Med.*, **134**, 886–906.

Swanson,J., Sparks,E., Zeligs,B., Siam,H.A. and Parrott,C. (1974) Studies on gonococcus infection. V. Observations on *in vitro* interactions of gonococci and human neutrophils. *Infect. Immun.*, **10**, 633–644.

Swanson,J., Sparks,E., Young,D. and King,G. (1975) Studies on gonococcus infection. X. Pili and leukocyte association factor as mediators of interactions between gonococci and eukaryotic cells *in vitro*. *Infect. Immun.*, **11**, 1352–1361.

Thomas,D.W., Hill,J.C. and Tyeryar,F.J., Jr. (1973) Interaction of gonococci with phagocytic leukocytes from men and mice. *Infect. Immun.*, **8**, 98–104.

Thongthai,C. and Sawyer,W.D. (1973) Studies on the virulence of *Neisseria gonorrhoeae*. I. Relation of colony morphology and resistance to phagocytosis by polymorphonuclear leukocytes. *Infect. Immun.*, **7**, 373–379.

Tramont,E.C. (1976) Specificity of inhibition of epithelial cell adhesion of *Neisseria gonorrhoeae*. *Infect. Immun.*, **14**, 593–595.

Tramont,E.C., Hodge,W.C., Gilbreath,M.J. and Ciak,J. (1979) Differences in attachment antigens of gonococci in reinfection. *J. Lab. Clin. Med.*, **93**, 730–735.

Trust,T.J., Lambden,P.R. and Watt,P.J. (1980) The cohesive properties of variants of *Neisseria gonorrhoeae* strain P9: specific pilus-mediated and non-specific interactions. *J. Gen. Microbiol.*, **119**, 179–187.

Virji,M. and Everson,J.S. (1981) Comparative virulence of opacity variants of *Neisseria gonorrhoeae* strain P9. *Infect. Immun.*, **31**, 965–970.

Virji,M. and Heckels,J.E. (1983) Antigenic cross-reactivity of *Neisseria* pili: investigations with type- and species-specific monoclonal antibodies. *J. Gen. Microbiol.*, **129**, 2761–2768.

Virji,M. and Heckels,J.E. (1984) The role of common and type-specific pilus antigenic domains in adhesion and virulence of gonococci for human epithelial cells. *J. Gen. Microbiol.*, **130**, 1089–1095.

Virji,M. and Heckels,J.E. (1985) Role of anti-pilus antibodies in host defence against gonococcal infection studied with monoclonal anti-pilus antibodies. *Infect. Immun.*, **49**, 621–628.

Virji,M. and Heckels,J.E. (1986) The effect of protein II and pili on the interaction of *Neisseria gonorrhoeae* with human polymorphonuclear leukocytes. *J. Gen. Microbiol.*, **132**, 503–512.

Virji,M., Everson,J.S. and Lambden,P.R. (1982) Effect of anti-pilus antisera on virulence of variants of *Neisseria gonorrhoeae* for cultured epithelial cells. *J. Gen. Microbiol.*, **128**, 1095–1100.

Virji,M., Heckels,J.E. and Watt,P.J. (1983) Monoclonal antibodies to gonoccal pili: studies on antigenic determinants of pili from variants of strain P9. *J. Gen. Microbiol.*, **129**, 1965–1973.

Walstad,D.L., Guymon,L.F. and Sparling,P.F. (1977) Altered outer membrane protein in different colonial types of *Neisseria gonorrhoeae*. *J. Bacteriol.*, **129**, 1623–1627.

Ward,M.E., Watt,P.J. and Robertson,J.N. (1974) The human fallopian tube: a laboratory model for gonococcal infection. *J. Infect. Dis.*, **129**, 650–659.

Winstanley,F.P., Blackwell,C.C., Tan,E.L., Patel,P.V., Parsons,N.J., Martin,P.M.V. and Smith,H. (1984) Alteration of pyocin-sensitivity pattern of *Neisseria gonorrhoeae* is associated with induced resistance to killing by human serum. *J. Gen. Microbiol.*, **130**, 1303–1306.

Witt,K., Veale,D.R., Finch,H., Penn,C.W., Sen,D. and Smith,H. (1976) Resistance of *Neisseria gonorrhoeae* grown *in vivo* to ingestion and digestion by phagocytes of human blood. *J. Gen. Microbiol.*, **96**, 341–350.

Zak,K., Diaz,J.-L., Jackson,D. and Heckels,J.E. (1984) Antigenic variation during infection with *Neisseria gonorrhoeae*: detection of antibodies to surface proteins in sera of patients with gonorrhea. *J. Infect. Dis.*, **149**, 166–173.

CHAPTER 7

Fimbrial variation in *Escherichia coli*

CYRIL J.SMYTH

Department of Microbiology, Moyne Institute, Trinity College, Dublin 2, Republic of Ireland

Introduction

Thirty years have now passed since the recognition by Duguid and co-workers that haemagglutination caused by strains of *Escherichia coli* could be related to the presence of fimbriae on these bacteria (Duguid *et al.*, 1955). The demonstration of specific types of fimbriae on enterotoxigenic strains of *E. coli* of human and animal origin and on isolates from urinary tract infections has led to intensive research into many aspects of fimbriae in relation to their role as adhesins to eukaryotic cells in infections at mucosal surfaces. In recent years advances in our knowledge of *E. coli* fimbriae have been made in several areas, in particular: (i) the development of techniques for rapid quantitative isolation and purification of fimbriae; (ii) physical and chemical characterization of isolated purified fimbriae; (iii) characterization of the nature and structure of cell surface receptors for fimbriae; and (iv) the genetic basis of expression and antigenic variation in *E. coli* through cloning techniques, restriction enzyme analysis of cloned DNA, transposon mutagenesis and nucleotide sequencing.

Several excellent reviews on the fimbriae of *E. coli* have been published in the last five years dealing with a number of general and specific aspects of their biology (Duguid and Old, 1980; Pearce and Buchanan, 1980; Gaastra and de Graaf, 1982; Ørskov and Ørskov, 1983; Levine *et al.*, 1983a; Jones and Isaacson, 1984; Kétyi, 1984; Wadström and Trust, 1984; Wadström *et al.*, 1984; Duguid, 1985; Freer, 1985; Klemm, 1985; Korhonen *et al.*, 1985a; Mooi and de Graaf, 1985; Parry and Rooke, 1985; Smith *et al.*, 1985; Uhlin *et al.*, 1985a). This chapter will deal specifically with data related to antigenic variation, phase variation and structural functional variation in *E. coli* fimbriae.

Antigenic variation in *E. coli* fimbriae

K88 antigen

The K88 antigen is present on many isolates of enterotoxigenic *E. coli* from piglets with diarrhoeal disease and was the first fimbrial antigen shown to be a virulence determinant (Ørskov *et al.*, 1961). Three antigenically different variants have been described, K88ab and K88ac (Ørskov *et al.*, 1964) and K88ad (Guinée and Jansen, 1979), the latter having emerged probably as a result of the selective pressures imposed by vaccination of pigs in Holland with vaccines containing the K88ab and K88ac antigens.

The antigenic factor common to all three variants is termed *a*, while the variable antigenic factors have been designated *b, c* and *d*. Additional minor serological variations in the K88ac antigen (Ørskov *et al.*, 1964) and K88ad antigen (Guinée and Jansen, 1979) have been observed, but their significance epidemiologically has not been assessed and they are not taken into account in routine diagnostic serotyping. Two gene subtypes of the K88ab antigen, termed K88ab1 and K88ab2, have been described recently on the basis of nucleotide sequencing data (Dykes *et al.*, 1985).

Irrespective of the antigenic variant studied, subunits of K88 antigen comprise 264 amino acids and have molecular weights of 25 000 – 27 500. The complete primary sequences of subunits of each of the variants have been determined by protein sequencing, nucleotide sequencing or both (Klemm, 1981; Gaastra *et al.*, 1981, 1983; Dykes *et al.*, 1985). The N-terminal 27 amino acids and the C-terminal 37 amino acids are identical in subunits of the K88 antigen variants. These regions have thus been proposed to play a role in stabilization of the polymeric structure of the fimbriae (Gaastra *et al.*, 1979; Klemm, 1981; Dykes *et al.*, 1985).

The K88ab1 and K88ab2 gene subtypes differed with respect to three base changes affecting amino acid residues at positions 133, 165 and 169, resulting in valine to alanine, threonine to asparagine and alanine to serine changes, respectively, in the K88ab2 protein (Dykes *et al.*, 1985). Nineteen base differences between the DNA sequences for the K88ab1 and K88ac variants resulted in 17 amino acid changes; in addition, the insertion of three bases in the K88ac DNA sequence relative to that for K88ab1 added an extra lysine at position 105 and the deletion of nine bases in the K88ac DNA sequence resulted in the removal of amino acid residues 165 – 167 of the K88ab1 subunit (*Table 1*). There were 37 base differences between the DNA sequences for K88ab1 and K88ad, resulting in 32 changes in the amino acid sequence of the K88ad subunit relative to that for K88ab1 subunit, but there were no deletions or insertions in K88ad DNA sequence relative to that of the K88ab1 subunit. Nine of the 17 amino acid changes in the amino acid sequence of the K88ac subunit relative to that of the K88ab1 subunit were also found in the K88ad subunit (Dykes *et al.*, 1985).

Apart from the gene subtypes of K88ab, the data of Dykes *et al.* (1985) and Gaastra *et al.* (1983) suggest different gene subtypes of K88ad, as there were 21 base differences in the respective DNA sequences which resulted in 11 amino acid changes. Such changes may be responsible for the slight serological variations beween the K88ad antigen in different strains that were originally described by Guinée and Jansen (1979).

Hydrophilicity profile analysis by the method of Hopp and Woods (1981) has been used to predict the antigenic determinants of the K88 subunit proteins from their amino acid sequences (Klemm, 1981; Klemm and Mikkelsen, 1982; Dykes *et al.*, 1985). Such analyses have also been supplemented by genetic experiments involving the reshuffling of certain parts of the gene encoding the K88 subunit protein among the different antigenic variants (Gaastra *et al.*, 1983). The fact that four of the six clusters of amino acid changes identified between variants coincided with regions which might be expected to be antigenic on the basis of Hopp and Woods' hydrophilicity analyses suggested that these regions probably contributed to the antigenic properties of the K88 antigen variants (*Table 1*). Amino acid sequences which might correspond to antigenic determinants associated with the *a, b, c* and *d* factors have been predicted (Klemm and Mikkelsen, 1982; Gaastra *et al.*, 1983; Dykes *et al.*, 1985).

Table 1. Differences in the amino acid sequences of antigenic variants of K88 antigen[a].

K88 antigen variant	*Amino acid sequences between indicated numbered residues for K88ab*													*Predicted antigenicity*
	73									83				
ab[b]	Val	Ser	Gly	Gly	Val	Asp	Gly	Ile	Pro	Gln	Ile			−
ac		Thr								His				−
ad		Thr	Ser							His				−
	93											104		
ab	Val	Lys	Leu	Arg	Asn	Thr	Asp	Gly	Glu	Thr	Asn	Lys	−[c]	++
ac		Val				Pro							Lys	++
ad		Glu				Pro					Glu			++
	130								138					
ab	Tyr	Ala	Gly	Val	Phe	Gly	Lys	Gly	Gly					−
ac					Leu		Arg							−
ad				Ala	Leu		Arg							−
	146											157		
ab	Leu	Phe	Ser	Leu	Phe	Ala	Asp	Gly	Leu	Arg	Ala	Ile		++
ac		Leu								Ser	Ser			−
ad		Met					Glu		Ser	His				−
	162												174	
ab	Leu	Thr	Thr	Thr	Val	Ser	Gly	Ala	Ala	Leu	Thr	Ser	Gly	−
ac		Pro	Arg	−[c]	−	−		Ser	Glu		Ser	Ala		++
ad		Pro		Asn		Lys	Asn	Ser	Glu		Lys	Gly		++
	212								220					
ab	Ser	Tyr	Arg	Glu	Asp	Met	Glu	Tyr	Thr					++
ac					Asn									++
ad		Phe	Asn		Asn		Ala							−

[a]Adapted from Dykes *et al.* (1985).
[b]Gene type K88ab1.
[c]−, deleted in sequence with respect to sequence of subunit(s) of other antigenic types of K88 antigen.

Apart from differences in antigenic determinants, the K88ab, K88ac and K88ad variants have been shown to possess different adhesive specificities for enterocytes derived from different pigs (Bijlsma *et al.*, 1981, 1982). Five porcine phenotypes were distinguished; one was receptive to the three variants, three were receptive to one or two variants, and one was non-receptive to all three variants. Data from studies of adhesion to brush border epithelia suggested that the receptors for K88ab, K88ac and K88ad antigens might be slightly different. Indeed, Parry and Porter (1978) had earlier shown that bacteria bearing the K88ac antigen did not agglutinate chicken erythrocytes whereas those bearing the K88ab antigen did.

It has been suggested that the antigenic variation of the K88 antigen has resulted from two possible selective pressures, namely, (i) the use of vaccines containing K88 antigen and (ii) the occurrence of pigs resistant to intestinal colonization by a particular serotypic variant because of altered receptors (Gaastra and de Graaf, 1982; Parry and Rooke, 1985). Indeed, in some countries, the K88-positive *E. coli* isolated from infected piglets now commonly possess either the K88ac or K88ad antigen while the K88ab antigen is found less frequently (Gaastra and de Graaf, 1982).

One interesting observation from the data of Dykes *et al.* (1985) is that every base difference noted in the DNA sequences for the K88ab, K88ac and K88ad subunits was associated with an amino acid sequence change, i.e. there were no silent mutations, supporting the view that positive selection pressures have been operative. However, it has not proved possible to predict the precise evolutionary relationship(s) between the K88 variants on the basis of the nucleotide sequence analyses. The observation of gene types for K88ab and possibly K88ad subunits suggests that there may be many intermediate gene subtypes between antigenic variants. Indeed, the data of Dykes *et al.* (1985) suggested that K88ab2 antigen was intermediate between K88ab1 and K88ad. Thus, the variable antigenic determinants, *b, c* and *d*, which are reflections of the differences in the DNA sequences for subunits, and the slight changes in biological properties of K88 fimbrial variants related to adhesiveness and receptor recognition may have evolved through genetic drift, selective pressures being the immune response of the host and changes in receptor structure or fit.

CS fimbriae

CFA/II. Evans and Evans (1978) described a fimbrial adhesin on enterotoxigenic *E. coli* of human origin which belonged to O-antigen serovars O6, O8, O80 and O85. Subsequently, this colonization factor antigen, termed CFA/II, was shown to be composed of three serologically distinct components which were designated coli-surface antigens (CS1, CS2 and CS3) (Smyth, 1982; Cravioto *et al.*, 1982). The CS1, CS2 and CS3 antigens were later identified as fimbriae (Mullany *et al.*, 1983; Smyth, 1984; Levine *et al.*, 1984). CS1 and CS2 fimbriae have a rigid, rod-like morphology with a diameter of 6–7 nm, whereas CS3 fimbriae are flexible wiry structures with a diameter of 2–3 nm. These fimbriae also differ with respect to the molecular weights of their subunits, and the haemagglutination patterns they confer on bacterial strains (Smyth, 1982; Cravioto *et al.*, 1982; Faris *et al.*, 1982; Levine *et al.*, 1984).

Strains expressing these fimbrial antigens under appropriate growth conditions may produce CS1 and CS3, CS2 and CS3, CS2 only or CS3 only (Smyth, 1982; Cravioto *et al.*, 1982; Mullany *et al.*, 1983; Smith *et al.*, 1983; Scotland *et al.*, 1985; Echeverria *et al.*, 1986). Strains bearing only CS3 fimbriae showed good adhesion to human small intestinal enterocytes, as did strains bearing both CS1 and CS3 fimbriae or CS2 and CS3 fimbriae (Knutton *et al.*, 1984a, 1984b, 1985). However, the relative roles of the latter two types of fimbriae in the adhesion of the dual fimbriate strains remains unclear. Expression of the CS antigens has been shown to be plasmid-mediated (Peñaranda *et al.*, 1980, 1983; Mullany *et al.*, 1983; Smith *et al.*, 1983; Boylan and Smyth, 1985; Echeverria *et al.*, 1986), although the plasmids associated with this property in different strains appear to be quite heterogeneous, with molecular weights ranging from 32 to 115 $\times$ 10^6. These plasmids generally also possess the genes for production of heat-labile (LT) and heat-stabile (ST) enterotoxins, although some only code for ST (Mullany *et al.*, 1983; Peñaranda *et al.*, 1980, 1983; Smith *et al.*, 1983; Echeverria *et al.*, 1986).

The CS fimbriae phenotype of a bacterial strain depends on the properties of the host bacterium although the nature of these remains unknown, but the serotype and biotype of the host are phenotypic markers of predictive value (*Table 2*). Mobilization exper-

Table 2. Expression of CS1, CS2 and CS3 fimbriae in naturally occurring enterotoxigenic *E. coli* and in wild-type or laboratory strains into which a CS fimbriae-associated plasmid has been mobilized or transformed[a].

Category of E. coli	*Serotype*	*Biotype*	*CS fimbriae expressed*
Natural wild-type hosts with CS fimbriae-associated plasmids[b]	O6:K15:H16 or H-	A	CS1 + CS3
	O6:K15:H16 or H-	B,C,F	CS2 + CS3; CS2 only
	O8:K40:H9 O8:H9 or H-	–	CS3 only
	O78:H12	–	CS3 only
	O80:H9	–	CS3 only
	O85:H7	–	CS3 only
	O115:H51	–	CS3 only
	O139:H28	–	CS3 only; CS1 + CS3
	O168:H16	–	CS3 only
Wild-type laboratory recipients of CS fimbriae-associated plasmids[c]	O6:H- O6:K14:H- O6:H1 O6:H13:H1 O6:K53 or K93:H7 O6:H10 O6:K15:H31	–	CS3 only
	O25:H42	–	CS3 only
	O78:H-	–	CS3 only
	O156:H-	–	CS3 only
Laboratory K-12 recipients of CS fimbriae-associated plasmids: strains J53, DU1000; DS410; C600; 14R519; 21R868; RR-1	–	–	CS3 only

[a]Data compiled from Peñaranda *et al.* (1980, 1983); Cravioto *et al.* (1982); Mullany *et al.* (1983); Smith *et al.* (1983); Smyth (1984); Boylan and Smyth, 1985; Scotland *et al.* (1985); M.Boylan and C.J.Smyth, unpublished data.
[b]In addition, enterotoxigenic *E. coli* isolates of serotypes O8:H?, O9:H2, O rough:H16, O?:H16, O128:H? have been described as CFA/II-positive (Thomas and Rowe, 1982; Levine *et al.* (1983b).
[c]Spontaneous CS fimbriae-negative variants of natural wild-type hosts when used as recipients for CS fimbriae-associated plasmids regain the same phenotype as the parent.

iments have only demonstrated that a single plasmid mediates expression of all three antigens, but have not revealed the mechanism underlying the host-related phenotypic expression of specific CS antigens (Mullany *et al.*, 1983; Smith *et al.*, 1983; Boylan and Smyth, 1985). With the exception of one wild-type isolate of enterotoxigenic *E. coli* of serotype O139:H28 (Scotland *et al.*, 1985), only enterotoxigenic *E. coli* of serotype O6:K15:H16 or H- belonging to definable biotypes can produce CS1 or CS2 fimbriae, which are mutally exclusive (*Table 2*). Enterotoxigenic *E. coli* of serotype O6:K15:H16 and of biotype C (rhamnose-positive; Scotland *et al.*, 1977) which have been reported to produce CS2 fimbriae but lack CS3 fimbriae (Smyth, 1982; Cravioto *et al.*, 1982) probably harbour a mutant plasmid or a plasmid with a deletion, as a plasmidless CS2 fimbriae-negative variant of such a strain was capable of expressing both CS2 and CS3 fimbriae when other plasmids were mobilized into it from either a CS1 and CS3 fimbriae- or CS2 and CS3 fimbriae-positive strain (Boylan and Smyth, 1985).

All other natural enterotoxigenic *E. coli* hosts of a CS antigen-associated plasmid, irrespective of O serovar or serotype, produce only CS3 fimbriae. Fourteen O:H antigen combinations have been described among serovar O6 *E. coli* isolates from various sources (Bettelheim, 1978). The number of permutations with K antigens included is unknown. Serovar O6 *E. coli*, other than those of serotypes O6:K15:H16 or H-, when used as recipients for mobilized plasmids only expressed CS3 fimbriae (Mullany *et al.*, 1983; Boylan and Smyth, 1985), including one strain of serotype O6:K15:H31 (Boylan and Smyth, unpublished data; Hacker *et al.*, 1985). Spontaneous CS3 fimbriae-negative variants of enterotoxigenic *E. coli* of O serovars other than O6 only express CS3 when used as recipients for mobilized plasmids (Mullany *et al.*, 1983; Boylan and Smyth, 1985; Scotland *et al.*, 1985).

A vexing phenomenon noted through mobilization studies is that in all *E. coli* K-12 hosts used to date any CS antigen-associated plasmid, irrespective of the serotype and biotype of the wild-type donor, only mediates expression of CS3 fimbriae (*Table 2*). CS1 and CS2 fimbriae are the only plasmid-mediated adhesins so far described in detail on enterotoxigenic *E. coli* which are not expressed in K-12 recipients. This phenonmenon has had a dramatic effect on DNA cloning experiments, as the testing of recombinant plasmids for their ability to mediate expression of CS1 or CS2 fimbriae or structural subunits has only been possible after their mobilization or transformation into CS fimbriae-negative, plasmidless variants of wild-type hosts of serotype O6:K15:H16 or H- and of appropriate biotype (*Table 2*).

Molecular cloning of the CS3 determinant has been achieved independently by three groups (Manning *et al.*, 1985; Boylan and Smyth, 1985; Boylan *et al.*, 1986; H.R. Smith, Central Publich Health Laboratory, London, unpublished data). A comparison of two of the restriction endonuclease maps of cloned *Hind*III fragments encoding the genes required for expression of CS3 fimbriae is shown in *Figure 1*. The minimum

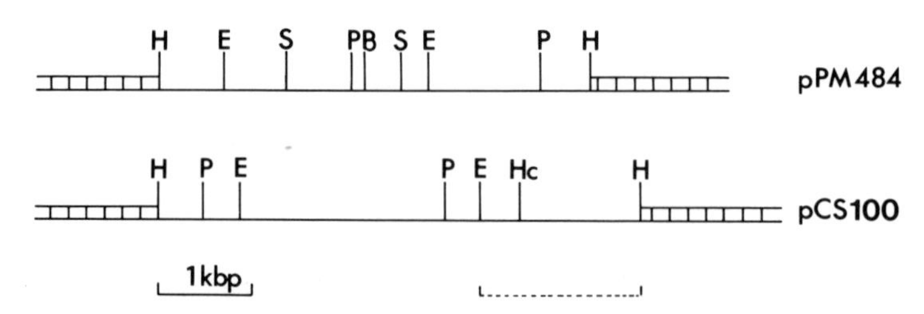

Fig. 1. Restriction endonuclease maps of DNA inserts in recombinant plasmids pPM484 and pCS100 encoding determinants for expression of CS3 fimbriae. Both recombinant plamids were constructed using vector plasmid pBR322. pPM484 contains a *Hind*III fragment of 4.6 kbp derived from the CS fimbriae-associated plasmid of strain PB176 (Evans and Evans, 1978; Manning *et al.*, 1985). pCS100 contains a *Hind*III fragment of 5.1 kbp derived from plasmid pCS001 obtained from strain E90a (Smyth, 1982; Boylan and Smyth, 1985; Boylan *et al.*, 1986). The hatched ends represent DNA of the cloning vector pBR322 and the thin black lines the DNA inserts of CS fimbriae-associated plasmid DNA. Restriction endonuclease cleavage sites are indicated by vertical lines: H, *Hind*III; E, *Eco*RI; S, *Sca*I; P, *Pst*I; B, *Bgl*II; Hc, *Hinc*II. Scale marker is 1 kbp. The dashed line indicates the 1.8-kbp DNA fragment of pCS100 in which the gene for the structural subunit of CS3 fimbriae was found; this was located by transposon mutagenesis and subcloning of DNA fragments in conjunction with minicell analysis for gene products.

regions necessary for expression have been further defined by transposon mutagenesis coupled with analysis of plasmin-encoded proteins in minicells (Manning *et al.*, 1985; Boylan *et al.*, 1986). These investigations have revealed that the cloned DNA fragments determining CS3 fimbriae expression contain clusters of genes encoding a number of proteins, the functions of which, other than that of the fimbrial subunit, are unknown. However, the genetic complexity of the CS3 fimbrial gene cluster probably resembles that described for other fimbrial antigens, whether plasmid- or chromosome-encoded (Mooi and de Graaf, 1985; Klemm, 1985; Uhlin *et al.*, 1985a). The genes mediating expression of CS1 and CS2 antigens have been cloned separately or together (H.R.Smith, unpublished data; M.Boylan, D.C.Coleman and C.J.Smyth, unpublished data).

However, although molecular cloning has confirmed that expression of CS3 fimbriae is associated with a cluster of genes on a non-autotransferring plasmid, including the gene for the structural subunit of the fimbriae, the host-specific expression of CS1 and CS2 fimbriae remains enigmatic. Several hypothetical explanations of this phenomenon are consistent with known data. Perhaps the simplest is that a gene (or genes) essential for expression of these fimbriae is only present in the chromosome of bacterial hosts of serotype O6:K15:H16 and of the appropriate biotypes, and that this gene is absent in alternative natural or laboratory host bacteria. The CS fimbriae-associated plasmid might contain either a regulatory sequence for CS1 and CS2 fimbriae which turns on otherwise silent genes in the chromosome of the serotype O6:K15:H16 host, or CS1 and CS2 fimbrial operons lacking regulatory sequences which are found only in these restricted hosts. Alternatively, the serotype O6:K15:H16 hosts may provide complementary genes required for the biogenesis of these fimbriae, the structural subunits of these fimbriae being plasmid encoded. The possibility also exists that the expression of CS1 and CS2 fimbriae is controlled by site-specific inversion of a DNA segment containing a promoter in a manner akin to that for flagellar phase variation in *Salmonella* (Silverman *et al.*, 1979; Kutsukake and Iino, 1980; Silverman and Simon, 1980; Zieg and Simon, 1980; Scott and Simon, 1982). This may involve specific factors or catalysts contributed by bacteria belonging to the limited host range in which these fimbriae are expressed.

Whatever the explanation, the CS1/CS2/CS3 fimbrial system surely holds many unique features. The most immediate aim should be to establish whether the structural genes for subunits of the CS1 and CS2 fimbriae are plasmid- or chromosome-encoded. Since a partial amino acid sequence for the CS2 fimbrial subunit is now know (Klemm *et al.*, 1985a), it should be possible to design a cDNA probe to establish this.

E8775 fimbriae. Following the discovery of CFA/I (Evans *et al.*, 1975) and CFA/II (Evans and Evans, 1978), a fimbrial antigen which conferred the property of mannose-resistant agglutination of human and bovine erythrocytes on an enterotoxigenic *E. coli* of serotype O25: H42 was described by Thomas *et al.* (1982) and was designated E8775 after the strain on which it was first observed. A survey of strains of enterotoxigenic *E. coli* isolated in Thailand, Bangladesh and Japan from cases of traveller's diarrhoea showed that about 5% of these strains possessed the E8775 fimbrial antigen, and that these belonged to O-serovars O25, O115 and O167 (Thomas and Rowe, 1982).

Recently, however, the nature of this fimbrial antigen was shown to be more complex, in a manner somewhat analogous to that for CFA/II which had been shown to comprise three antigenically distinct fimbriae (*vide supra*) (Thomas *et al.*, 1985). Surface extracts of strains of O-serovar O25 were shown to contain two antigenic components termed coli surface antigens CS4 and CS6, whereas surface extracts of strains of serovars O115 and O167 revealed antigens CS5 and CS6. By immune electron microscopy, CS4 and CS5 were shown to be rigid rod-like fimbriae about 7 nm in diameter. The morphological nature of CS6 was not revealed (Thomas *et al.*, 1985). However, given that the CS3 antigen was not shown to be fimbriate for several years after its discovery, despite extensive and painstaking attempts using a variety of strains for transmission electron microscopy and immune electron microscopy (Mullany *et al.*, 1983; Smyth, 1984; Levine *et al.*, 1984), CS6 may yet prove to be a fimbrial antigen of the fine flexible type.

A plasmid mediating expression of CS5 and CS6 has been transferred from an O-serovar O167 strain into *E. coli* K-12 (unpublished data cited by Thomas *et al.*, 1985). However, the basis for expression of CS4 and CS6 is unclear although it may also be plasmid-mediated. Whether a single plasmid mediates expression of CS4 and CS6 and of CS5 and CS6 in strains of different serotypes in a manner analogous to CS1/CS2 and CS3 expression remains to be established, although suggestive parallel phenotypic features seem to apply.

Type 1 or common type fimbriae

Type 1 fimbriae occur widely among the enterobacteria. They share a common rigid rod-like morphology, and their adhesive properties are susceptible to complete inhibition by D-mannose and a few analogues of this sugar (Old, 1972; Duguid and Old, 1980). Type 1 fimbriae are produced by as many as 70% of *E. coli* strains of clinical or environmental origin (Ørskov *et al.*, 1982a).

Type 1 fimbriate bacteria adhere to a broad range of mammalian host cells because of the ubiquitous occurrence of mannose in host cell membranes (Duguid, 1985; Duguid and Old, 1980; Parry and Rooke, 1985). However, the physiological function of type 1 fimbriae and their role, if any, in the pathogenesis of *E. coli* infections remains unclear (Duguid, 1985; Duguid and Old, 1980; Korhonen *et al.*, 1985a; Parry and Rooke, 1985).

Several groups have considered that type 1 fimbriae do not play a role in the adhesion of *E. coli* to human uroepithelial cells (Svanborg-Edén and Hansson, 1978; Källenius and Möllby, 1979; Ørskov *et al.*, 1980b; Korhonen *et al.*, 1981). Tamm–Horsfall glycoprotein or uromucoid present in urinary mucus on the surface of uroepithelial cells lining the bladder has been shown to trap type 1 fimbriate bacteria in a mannose-specific manner (Ørskov *et al.*, 1980a, 1980b; Parry and Rooke, 1985). Binding to urinary mucus may prevent bacterial adhesion to underlying epithelial cells and lead to elimination of type 1 fimbriate bacteria. However, bacterial attachment to or entrapment in urinary mucus could have a role in persistence of bacteria in the bladder.

Abraham *et al.* (1985) showed that passive intraperitoneal administration of monoclonal antibodies directed against antigenic epitopes inherent in the supramolecular structure of type 1 fimbriae or against D-mannose residues of their complementary receptors conferred protection on mice against retrograde colonization with type 1 fimbriate *E. coli* instilled into their bladders. Interestingly, Kuriyama and Silverblatt (1986) have shown that, whereas human polymorphonuclear leukocytes (PMN) ingested type 1 fimbriate *E. coli* even in the absence of antibody, complement or other serum opsonins,

Tamm–Horsfall protein at physiological concentrations interfered with serum-dependent ingestion by forming a pseudocapsule around bacteria bearing type 1 fimbriae and suggested that this might contribute to the virulence of *E. coli* in the bladder.

Ørskov *et al.* (1980b) suggested that adhesion to slime in the large intestine might be mediated by type 1 fimbriae and that these mannose-sensitive fimbriae might function as a colonization factor of the normal flora. However, it is unclear if type 1 fimbriae act as an adhesive factor mediating attachment of enterotoxigenic *E. coli* to the small intestinal mucosa. To *et al.* (1984) demonstrated that vaccination of dams with a type 1 fimbrial vaccine did not significantly protect suckling piglets from challenge with the parent *fim*$^+$, K99 antigen-positive enterotoxigenic *E. coli*. Moreover, no evidence for the production of type 1 fimbriae *in vivo* was obtained. In addition, Levine *et al.* (1982) have interpreted the failure of human volunteers to develop rises in antibody titres to type 1 fimbriae after challenge with enterotoxigenic *E. coli* as evidence that type 1 fimbriae may not be expressed in the course of human infections.

In contrast, Jayappa *et al.* (1985) showed that type 1 fimbriate bacteria readily attached to the small intestines of colostrum-deprived newborn piglets, and that a purified type 1 fimbrial vaccine administered to dams conferred significant protection against colibacillosis in newborn pigs following challenge with enterotoxigenic *E. coli* expressing type 1 fimbriae. The discrepancies between the above-mentioned studies may be related to the differences in adhesive properties for intestinal membranes, haemagglutination patterns and surface charge and hydrophobicity observed among type 1 fimbriate *E. coli* by Sherman *et al.* (1985), who suggested that some type 1 fimbriae might contribute to intestinal adherence.

Although it has previously been accepted that type 1 fimbriae within an enterobacterial species were serologically related using whole-cell antisera absorbed with non-fimbriate variants (Duguid and Campbell, 1969; Nowotarska and Mulczyk, 1977), it is now appreciated that either (i) type 1 fimbriae may possess major and minor antigenic determinants as revealed by agglutination tests with bacterial suspensions (Duguid, 1985) or (ii) single strains may express two species of antigenically distinct type 1 fimbriae as demonstrated using crossed immunoelectrophoresis and immune electron microscopy (Klemm *et al.*, 1982; Salit *et al.*, 1983).

The observations of Rhen *et al.* (1983b, 1983d) on rapid phase variation of P-fimbrial antigens may be pertinent to this dichotomy in interpretation of results. Their findings with agar-grown *E. coli* demonstrated that different fimbriae did not apparently exist on a single cell, but rather that the bacterial population within a single colony consisted of a heterogeneous mixture of cells expressing different fimbriae. Separated populations of different fimbrial phenotype gave rise to colonies expressing the spectrum of fimbriae characteristic of the original strain (Rhen *et al.*, 1983b; Korhonen *et al.*, 1985a). Thus, the reported demonstration of strains with so-called major and minor type 1 fimbrial antigenic determinants may really reflect the relative proportions of bacteria bearing antigenically distinct fimbriae or fimbriae with some or little immunological cross-reactivity.

Although amino acid compositions, terminal sequences and physicochemical properties of highly purified type 1 fimbriae have been determined (Korhonen *et al.*, 1980; Eshdat *et al.*, 1981; Klemm *et al.*, 1982), no relationship between antigenic variation and amino acid sequence has yet been revealed. The nucleotide sequence for and the primary structure of the type 1 fimbrial subunit containing 158 amino acids (the *fimA* gene product) have been determined for at least one strain (Klemm, 1984, 1985). The N- and C-terminal amino acid sequences of three antigenically distinct but related type 1 fimbriae, termed 1A, 1B and 1C, found on uropathogenic strains of *E. coli* were highly conserved indicating that sequence divergence accounting for antigenic heterogeneity may be found in central parts of the primary structures of these antigenic variants (Klemm *et al.*, 1982; Ørskov *et al.*, 1982b). Type 1C fimbriae conferred adherence to buccal epithelial cells but, in contrast to type 1A fimbriae, did not confer adherence to uroepithelial cells or to urinary mucus and were non-haemagglutinating (Klemm *et al.*, 1982; Ørskov *et al.*, 1982b) and were probed newly with monoclonal antibodies (Schmitz *et al.*, 1986).

It is unknown how the type 1A, 1B and 1C fimbriae relate to the major type 1 fimbrial antigens 4, 5, 6 described by Duguid *et al.* (1979) and Duguid (1985). In addition to serologically distinct types of fimbriae, most type 1 fimbriate *E. coli* belonging to 84 O-serovars appeared to share a fimbrial antigen, termed 2, with strains of *Shigella flexneri* (Gillies and Duguid, 1958; Duguid, 1985). Thus, despite advances in the molecular biology and genetics of type 1 fimbriae (Klemm, 1985), research on antigenic types to explain variation is still in its infancy.

P-fimbriae

The past 10 years have seen intensive study of *E. coli* isolates from urinary tract infections and pyelonephritis. Many such isolates cause mannose-resistant haemagglutination of human erythrocytes and adhere to human uroepithelial or periurethral cells *in vitro* (Svanborg-Edén *et al.*, 1977; Källenius and Möllby, 1979; Källenius *et al.*, 1980a). The glycosphingolipids containing a terminal digalactoside, α-D-Gal*p*-(1-4) β-D-Gal*p*, comprising the P blood group antigens have been shown to act as the receptors for rigid, rod-like fimbriae on the surface of a high percentage of uropathogenic *E. coli* (Källenius *et al.*, 1980b; Leffler and Svanborg-Edén, 1980; Väisänen *et al.*, 1981; Svenson and Källenius, 1983; Svenson *et al.*, 1983). The fimbriae with digalactoside receptors are variously referred to as P-fimbriae, Pap fimbriae (*p*ili *a*ssociated with *p*yelonephritis), Gal−Gal-binding fimbriae, and F7 fimbriae (Klemm *et al.*, 1982; Korhonen *et al.*, 1982; Normark *et al.*, 1983; O'Hanley *et al.*,1983). A receptor-specific particle agglutination test has been developed for rapid identification of P-fimbriated *E. coli* (Svenson *et al.*, 1982) which occur more frequently among pyelonephritogenic isolates than on strains from cystitis, asymptomatic bacteriuria or faeces (Källenius *et al.*, 1981; Väisänen-Rhen *et al.*, 1984). *E. coli* strains of serotypes O1:K1:H7, O4:K12: H5, O6:K2:H1, O16:K2:H1, O16:K1:H6 and O18ac:K5:H7 commonly possess P-fimbriae (Väisänen-Rhen *et al.*, 1984).

Any agar-grown uropathogenic *E. coli* isolate may produce more than one antigenic variant of P-fimbriae. These have been variously described as KS71A and KS71B fimbriae (Rhen *et al.*, 1983a, 1983b), $F7_1$ and $F7_2$ fimbriae (Ørskov and Ørskov, 1983), and, F9 (Ørskov *et al.*, 1982b), F12 (Klemm *et al.*, 1983) and F13 fimbriae (Normark *et al.*, 1983; Uhlin *et al.*, 1985a). The P-fimbriae KS71A and KS71B cross-reacted antigenically despite being composed of subunits with apparently different molecular weights by sodium dodecyl sulphate polyacrylamide gel electrophoresis (SDS−PAGE)

(Rhen *et al.*, 1983a; Korhonen *et al.*, 1985a). In addition, strain KS71 produced so-called KS71C fimbriae which were immunologically distinct from type 1 fimbriae (KS71D) and P-fimbriae (KS71A and KS71B) and were non-haemagglutinating (Rhen *et al.*, 1983a, 1983c). It seems likely that multiple subunits with apparently different molecular weights, as revealed in purified fimbrial preparations from uropathogenic *E. coli* (e.g. Jann *et al.*, 1981; Korhonen *et al.*, 1982), were probably derived from antigenically different but morphologically indistinguishable fimbriae. However, the fractionation of the F7 fimbriae into two fimbrial antigens termed $F7_1$ and $F7_2$, as revealed by separate precipitation lines in crossed immunoelectrophoresis, emphasizes that the observation of an apparent single polypeptide band in SDS−PAGE should not be interpreted as proof of the antigenic homogeneity of fimbriae (Ørskov and Ørskov, 1983; Korhonen *et al.*, 1985a).

The number of antigenic types of P-fimbriae is unknown. Recently, Hanley *et al.* (1985) attempted to clarify the antigenic diversity of P-fimbriae derived from invasive strains of *E. coli* using Western blotting, an enzyme-linked immunosorbent inhibition assay and N-terminal protein sequencing. Very limited antigenic cross-reactivity was found between 14 strains using antisera raised to the P-fimbriae of four of these strains. The amino-terminal sequences of the P-fimbriae of the latter four strains were identical through the first 10 residues and essentially identical to the sequences published for $F7_1$, $F7_2$ and Pap fimbriae (Klemm, 1985).

De Ree *et al.* (1985a, 1985b) raised monoclonal antibodies (MAbs) against five different P-fimbriae, namely $F7_1$, $F7_2$, F9, F11 and Pap. Four different anti-$F7_1$ MAbs only recognized epitopes on the homologous fimbriae; of seven anti-$F7_2$ MAbs, two recognized epitopes on $F7_2$ fimbriae only, two recognized epitopes on both $F7_2$ and F9 fimbriae, and three reacted only with epitopes on F9 fimbriae. Three anti-F9 MAbs were type-specific. Five MAbs to Pap fimbriae only recognized epitopes on homologous fimbriae, whereas another recognized an epitope on Pap, $F7_2$ and F11 fimbriae. Of six MAbs against F11 fimbriae, two recognized F11, Pap and $F7_2$ fimbriae and four recognized F11 fimbriae only. These findings are consistent with the observations of Hanley *et al.* (1985) using polyclonal antisera and with the striking homologies in amino acid sequences beween the Pap and $F7_2$ fimbrial variant subunits (Klemm, 1985), and go some way towards rationalization of the apparent discrepancy in observations of antigenic cross-reactivity between KS71A and KS71B P-fimbrial variants (Rhen *et al.*, 1983a, 1983b) and the lack of it between $F7_1$ and $F7_2$ P-fimbrial variants (Ørskov and Ørskov, 1983).

The overall P-fimbrial antigenic pattern appears to be related to the O-serovar of the strain. P-fimbriae on pyelonephritogenic *E. coli* belonging to O-serovars O2, O4 and O6 were analysed by immune precipitation (Pere *et al.*, 1985; Korhonen *et al.*, 1985a, 1985b, 1985c). The P-fimbriae of strains belonging to O-serovar O2 were serologically different from those of strains within O-serovars O4 and O6 which appeared to react with antisera to the KS71A and KS71B P-fimbrial antigenic variants. Most P-fimbriate strains had more than one P-fimbrial antigen which differed in apparent subunit molecular weight by SDS−PAGE as well as serological properties.

Thus, despite identical amino-terminal amino acid sequences, P-fimbriae appear to be antigenically heterogeneous. It remains to be seen whether these chromosomally encoded fimbriae will prove to be as antigenically complex as the chromosomally encoded fimbriae of *Neisseria gonorrhoeae* (So *et al.*, 1985a, 1985b).

The genetics of the biogenesis of P-fimbriae have been reviewed in depth (Uhlin *et al.*, 1985a). This digalactoside-specific fimbria is expressed from a multicistronic cluster of genes denoted as *pap* genes, of which nine have been identified, viz. *papA* to *papI* (Uhlin *et al.*, 1985a; Normark *et al.*, 1985), encoding products involved in regulation, assembly, secretion, compartmentalization and fimbria structure (see section on **Structural and functional aspects of fimbriae**). The complete DNA sequence data for subunits of antigenic variants of P-fimbriae are not yet available to allow antigenic differences to be interpreted in terms of amino acid sequences. However, antigenic variation is also related to phase variation involving switches in fimbrial subunit synthesis in individual cells (Rhen *et al.*, 1983d).

The structural genes for the KS71A and KS71B P-fimbriae have been cloned separately in *E. coli* K-12 from the uropathogenic *E. coli* strain KS71 of serotype O4:K12 (Rhen *et al.*, 1983c; Rhen, 1985), as have the $F7_1$ and $F7_2$ genes from strain C1212 (Ørskov *et al.*, 1982a, Ørskov and Ørskov, 1983; Low *et al.*, 1984). The gene clusters for KS71A and KS71B fimbriae have been shown to complement each other in *trans*, but not that for the immunologically distinct non-P KS71C fimbriae (Rhen *et al.*, 1985). A *trans*-acting regulatory element was located close to the KS71A structural subunit gene. In addition, copies of gene sequences homologous to the entire *pap* operon have been found to exist at multiple sites in the chromosomes of some strains (Hull *et al.*, 1985). Whether or not antigenic variation evolved by intragenic recombination events involving semi-variable and hypervariable segments of the *papA* structural gene in a manner analogous to that envisaged for gonococcal fimbriae (So *et al.*, 1985a, 1985b) can only be evaluated when the sequences of genes from a large number of strains have been determined.

Other fimbriae

On enterotoxigenic *E. coli* of human and animal origin a number of fimbrial adhesins have been described in addition to those dealt with above. None of these, to this author's knowledge, has been shown to exhibit antigenic heterogeneity or variation. Although the K99 antigen was demonsterated at one stage to be apparently composed of an anionic and a cationic component (Morris *et al.*, 1978), subsequent investigation demonstrated that these components were distinct adhesive antigens, namely the F41 and K99 fimbriae, respectively (Morris *et al.*, 1980, 1982, 1983). These fimbriae differ with respect to pI, the molecular weights of subunits and the haemagglutination properties conferred on enterotoxigenic *E. coli* (Mooi and de Graaf, 1985; Parry and Rooke, 1985).

Colonization factor antigen I (CFA/I) was the first to be described on enterotoxigenic *E. coli* of human origin (Evans *et al.*, 1975). Antisera raised to bacterial strains of differing serotype bearing these fimbriae, after absorption with CFA/I$^-$ derivatives of the respective immunogens, are fully and equally reactive with CFA/I fimbriae on heterologous strains (Gross *et al.*, 1978; Thomas and Rowe, 1982; H.R.Smith, personal communication). Some serotypes of *E. coli* commonly encountered among human enterotoxigenic strains do not appear to possess the fimbrial adhesins already described, e.g. strains of serotypes O148:H28 and O159:H4. Since these strains adhere to the intestinal mucosa, it seems likely that other, at present unrecognized, antigenically distinct fimbriae will eventually be revealed on such strains (Levine *et al.*, 1984).

Honda *et al.* (1984) have described a new and antigenically distinct colonization factor antigen, CFA/III, on a group of enterotoxigenic *E. coli* isolates of human origin which had been initially characterized by their heat-stable hydrophobicity (Honda *et al.*, 1983). The antigen was fimbrial in nature. Darfeuille *et al.* (1983) also described an antigen which they termed CFA/III on an *E. coli* of serotype O128:B12. However, properties of the strain on which this adhesin was described and other studies on adhesins present on enteropathogenic *E. coli* of this serotype (Scotland *et al.*, 1981; Guth *et al.*, 1985; Echeverria *et al.*, 1986) strongly suggest that the latter 'CFA/III' was in fact CFA/I. Indeed, plasmid analysis of the strain used by Darfeuille *et al.* (1983) has revealed the presence of a typical CFA/I-associated plasmid which hybridized with a DNA probe for CFA/I antigen (P.Manning, University of Adelaide, personal communication).

Enterotoxigenic *E. coli* isolates from infantile diarrhoea have also been shown to bind to fibronectin, a glycoprotein found in the extracellular matrix of loose connective tissue and in plasma (Fröman *et al.*, 1984). Fibronectin binding has also been shown to be a common characteristic of *E. coli* isolates from bovine mastitis (Faris, 1985). The adherence of such strains to tissue culture fibroblasts or exfoliated cells was mediated by fimbriae which have been designated Fib fimbriae. These fimbriae appear to be distinct from others described on *E. coli* and are morphologically of the thin, wiry type with a diameter of 4 nm (Faris, 1985).

The search for adhesins on porcine isolates of enterotoxigenic *E. coli* which lacked K88 and K99 fimbriae led to the discovery of 987P fimbriae (Nagy *et al.*, 1976, 1977). A survey of the fimbrial types of porcine enterotoxigenic *E. coli* lacking K88 antigen revealed that about 75% of these possessed 987P fimbriae (Moon *et al.*, 1980). Strains of *E. coli* belonging to O-serovar O115 have been associated with severe catarrhal to haemorrhagic enteritis and septicaemia in calves and pigs (Ørskov and Ørskov, 1978). A new mannose-resistant haemagglutinating fimbrial antigen designated F165 has been found on non-enterotoxigenic strains of this serogroup isolated from piglets with diarrhoea (Fairbrother *et al.*, 1986). The role of this fimbria in infection remains to be established.

Two types of fimbriae other than P-fimbriae, KS71C fimbriae and type 1 fimbriae have been described on uropathogenic *E. coli* or on isolates from cases of *E. coli* meningitis and neonatal sepsis (Kusecek *et al.*, 1984), namely, M-fimbriae (Väisänen *et al.*, 1982) and S-fimbriae (Parkkinen *et al.*, 1983; Korhonen *et al.*, 1984, 1985b; Hacker *et al.*, 1985). The receptor specificities of the fimbriae described by Karch *et al.* (1985a) on an isolate of serotype O7:K1:H6 and by Väisänen-Rhen (1984) on O-serovar O75 isolates remain to be elucidated. The term X adhesin has been used for such mannose-resistant fimbriae and haemagglutinins (Väisänen-Rhen *et al.*, 1984).

Phase variation of *E. coli* fimbriae

The ability of a bacterium to adhere depends not only on its ability to synthesize a ligand such as a fimbria, but also on expression of the ligand in an accessible configuration on the bacterial surface. Synthesis and expression are under genetic and phenotypic control. Comparison of a parent strain with laboratory derivatives in terms of adhesive capacity and antigenicity has proved useful in elucidating the mechanism of adherence and the role of fimbriae in the pathogenesis of infection, despite occasional problems due to genotypic or phenotypic variants lacking or acquiring properties other than the one being tested by an investigator.

A pure culture of a bacterial species is often capable of spontaneously producing variants with easily distinguishable forms, properties or phenotypes. In the case of bacteria expressing plasmid-mediated fimbriae, spontaneous loss of the plasmid from an *E. coli* strain is associated with its irreversible conversion from a fimbriate to a non-fimbriate phenotype. In contrast, bacteria which are genotypically capable of producing either plasmid-mediated or chromosomally encoded fimbriae can also vary reversibly between fimbriate and non-fimbriate phenotypes, a phenomenon commonly referred to as fimbrial phase variation (Brinton, 1978; Brinton *et al.*, 1978).

In fimbrial phase variation, each phase is metastable, i.e. the daughter cells retain the phenotypic state of the mother cell in the absence of phase change. The existing phenotypic state can switch spontaneously to another phenotypic state and this change is reversible but does not necessarily occur at the same rate in both directions, i.e. fimbriate to non-fimbriate and vice versa. Moreover, the observation that a single strain may apparently express several different types of fimbriae has led to the realization that phase variation in fimbrial antigens can involve the synthesis of different fimbrial proteins by individual cells within a single bacterial population (Rhen *et al.*, 1983d; Korhonen *et al.*, 1985a).

Expression of plasmid-mediated fimbriae

Although growth medium composition can be readily demonstrated to affect fimbrial expression, the precise nature of factors involved and molecular mechanisms of gene regulation are largely unknown. For example, production of K99 fimbriae in strains containing wild-type K99-associated plasmids or the recombinant plasmid pFK99 (plasmid pBR322 into which was cloned an 8.2-kbp DNA insert containing the gene cluster for K99 fimbriae biogenesis) was repressed by L-alanine, but which gene or genes were regulated by L-alanine was not elucidated (de Graaf *et al.*, 1980; van Embden *et al.*, 1980). Glucose, pyruvate and arabinose have also been reported to repress K99 fimbriae biogenesis (Guinée *et al.*, 1977; Isaacson, 1980). Glucose-mediated repression appeared to be due to cAMP-dependent catabolite repression (Isaacson, 1980).

F41 fimbriae production has also been reported to be repressed by L-alanine and this has led to the suggestion that regulation of the production of both F41 and K99 fimbriae may have a similar basis (de Graaf and Roorda, 1982).

Recently, it has been demonstrated that production of both types of fimbriae only occurred during balanced growth with high biomass yields at specific growth rates (μ) higher than 0.04 h^{-1} corresponding to a mass doubling time of less than 17 h (van Verseveld *et al.*, 1985). These observations suggest that expression of fimbriae is only permitted at low concentrations of intracellular guanosine polyphosphates and cAMP, and is arrested at high levels of these regulatory nucleotides (Contrepois *et al.*, 1983; van Verseveld *et al.*, 1985). However, assembly rather than synthesis of fimbrial subunits appears to be restricted and cells may possess a large pool of unassembled subunits (Isaacson, 1983). Such pools of pre-existing subunits may be assembled into fimbriae when ingested cells meet favourable growth conditions in the small intestine leading to adherence and proliferation (van Verseveld *et al.*, 1985).

A feature common to plasmid-mediated fimbriae is thermoregulation of expression, which was first shown for K88 fimbriae (Ørskov *et al.*, 1961; Gaastra and de Graaf, 1982). Optimal expression of fimbriae occurs at 35−37°C with quantitative reduction below 30°C. Fimbriae are absent below 20°C. Such thermoregulatory effects have also been shown in *E. coli* K-12 hosts into which wild-type plasmids have been transferred or mobilized and in such hosts containing recombinant plasmids bearing the genes for K88 biogenesis (Isaacson, 1980; Mooi *et al.*, 1979).

The optimal expression of plasmid-mediated fimbriae at physiological temperatures, at which they have been unequivocally shown to play an adhesive role in the pathogenesis of intestinal infections, has an apparent logic. Since fimbrial biogenesis may account for 1−2% of total protein synthesis (Klemm, 1985), regulation of fimbrial biosynthesis under conditions which are non-physiological would permit considerable energy savings, thereby providing greater potential for growth and maintenance.

Genetic mapping of cloned DNA fragments encoding biogenesis of K88ab and K99 fimbriae has revealed the presence of small intracistronic regions with dyad symmetry between fimbrial subunit genes and genes immediately downstream, encoding polypeptides of 81 and 76 kd (p82 and p76) in the K88ab and K99 gene clusters, respectively (Mooi and de Graaf, 1985). The p81 and p76 polypeptides are located in the outer membrane and have been assigned putative functions in the anchorage of fimbriae and the transport of other polypeptides across the outer membrane (Mooi and de Graaf, 1985). Stem and loop structures can be formed by mRNA transcripts of the intracistronic regions with dyad symmetry. It has been suggested that, since the putative initiation codons of both the p81 and p76 proteins are located within these stem and loop structures, these regions may be involved in temperature-dependent regulation (Mooi and de Graaf, 1985). At low temperatures, the stems formed by the mRNA transcripts would be stabilized thereby preventing the translation of the p81 and p76 mRNAs. Other studies with the recombinant K88ab plasmid pFM205 suggest that temperature does not affect initiation of transcription of K88ab genes (Mooi *et al.*, 1979). Thus, the data to date appear to indicate that thermoregulation functions at the translational level for these plasmid-encoded fimbriae.

Growth of bacteria at different temperatures is known to result in changes in the degree of saturation of the phospholipid acyl chains in membranes so as to maintain membrane fluidity, so-called homeoviscous adaptation (Houslay and Stanley, 1982; McElhaney, 1982, 1984). Such changes in membrane phospholipid fatty acid composition may have as yet undefined effects either directly or indirectly on fimbrial biogenesis. Inhibition of a wide variety of membrane-associated enzymes and transport systems has been shown to accompany the formation of gel-state lipid in intact microbial cells and, in most instances, these activities are severely impaired at temperatures below the phase transition mid-point temperature (McElhaney, 1982). The transport of low molecular weight fimbrial subunits across the inner and outer membranes and their assembly into supramolecular structures, i.e. phenotypic expression as fimbriae, has been shown to involve several gene products. Membrane phospholipid fatty acid composition might affect secretion of fimbrial subunits, or of components which are involved in subunit transport across the outer membrane, or in their anchorage, or in the formation of transport channels in the outer membrane, as envisaged in proposed models for the biogenesis of K88 and K99 fimbriae (Mooi and de Graaf, 1985).

Expression of CFA/I fimbriae has been shown to involve two widely separated regions, designated 1 and 2, of the plasmid encoding CFA/I (Smith *et al.*, 1982; Willshaw *et al.*, 1985). Region 1 directed production of at least six independent polypeptides in *E. coli* minicells including the fimbrial subunit. Region 2 specified three polypeptides in minicells. A product encoded by region 2 was required for fimbrial assembly but not for subunit production, and may be involved in stabilizing the fimbrial subunit or converting it to a form that can be transported into the outer membrane (Willshaw *et al.*, 1985; Mooi and de Graaf, 1985).

Loss of CFA/I and CFA/II or CS fimbriae expression by human enterotoxigenic *E. coli* has been observed to occur occasionally without the apparent concomitant loss of a plasmid (Evans *et al.*, 1981; Levine *et al.*, 1983b; Echeverria *et al.*, 1986). Simultaneous loss of production of ST enterotoxin or of the ST gene did not always occur. The gene encoding ST production is known to be located on a transposable element (So *et al.*, 1979). Such insertion elements have been shown to be responsible for a variety of deletion, duplication, inversion and transposition events in bacteria. Transposition of the ST transposon may affect expression of adjacent genes, such as those encoding CFA/I fimbriae which are so located (Smith *et al.*, 1982, 1985; Willshaw *et al.*, 1985).

Expression of chromosomally encoded fimbriae

Serial subculture of *E. coli* strains in non-aerated, unshaken broth cultures has been known for 30 years to enhance expression of type 1 fimbriae, whilst growth on well-dried agar plates gives cultures with few or no type 1 fimbriate bacteria (Duguid and Old, 1980; Duguid, 1985). Spontaneous variation between fimbriate and non-fimbriate phases in *E. coli* can occur in either direction at a rate of 1 in 1000 to 1 in 10^4 bacteria per generation (Brinton, 1965, 1967). Since several passages under appropriate cultural conditions are required to complete a phase change, it appears that the effect of the growth conditions is to enhance outgrowth of spontaneously arising variant bacteria rather than directly to induce or repress fimbrial synthesis.

Type 1 fimbriae promote growth of fimbriate bacteria at the surface of the unshaken broth as a thin pellicle, where the supply of oxygen allows them to multiply more readily, possibly by virtue of fimbrial hydrophobicity which allows bacteria to adhere to each other and holds them at the air–water interface. Although type 1 fimbriae were thought to be subject to cAMP-dependent catabolite repression by carbohydrates (Saier *et al.*, 1978; Ofek *et al.*, 1981), the degree of fimbriation of a culture appears to be determined by alteration in the rate of phase variation rather than by catabolite repression (Eisenstein and Dodd, 1982).

Type 1 fimbria expression has been demonstrated to be subject to metastable regulation at the transcriptional level (Eisenstein, 1981). The genes involved in the biogenesis of type 1 fimbriae have been cloned (Hull *et al.*, 1981; Klemm *et al.*, 1985b) and the *fimA* gene encoding the structural subunit of type 1 fimbriae has been sequenced (Klemm, 1984). Using a constructed chromosomal *fimA–lacZ* gene fusion, the *fimA* gene was subject to metastable transcriptional regulation (Orndorff *et al.*, 1985). The genetic element (or elements) determining transcription of the *fimA* gene appeared to be closely linked to it (Orndorff and Falkow, 1984). In addition to the metastable regulation of the *fimA* gene, transcriptional regulation was also affected by the product of a gene adjacent to the *fimA* gene termed *hyp*. The gene product of *hyp* acted as a repressor of *fimA* transcription (Orndorff *et al.*, 1985). The *hyp* gene, however, did not appear to affect or effect metastable variation, but rather the level of transcription of the *fimA* gene in the 'on' or transcribed mode.

Investigation of the genetic regulation of P-fimbriae has also been greatly aided by the cloning of chromosomal genes involved in their production (see Uhlin *et al.*, 1985a for review). The construction of recombinant plasmids encoding type 1 fimbriae, non-haemagglutinating K71C fimbriae or P-fimbriae from a single urinary tract infection *E. coli* isolate showed that the genes involved in the biogenesis of these fimbriae were distinct (Hull *et al.*, 1981; Rhen *et al.*, 1983c, 1985; Rhen, 1985). Indeed, the structural genes for the P-fimbrial antigenic variants of strain KS71 (KS71A and KS71B) have been cloned separately into *E. coli* K-12, demonstrating that they are located at different sites on the chromosome (Rhen *et al.*, 1983c, 1985; Rhen, 1985). Characterization of the organization of the genes in cloned DNA allowing expression of P-fimbriae has indicated that there may be at least four functional promoters involved in transcriptional events (Uhlin *et al.*, 1985a). The *papA* gene codes for the P-fimbrial subunit polypeptide and the *papB* and *papI* gene products appear to be involved in transcriptional regulation of the *papA* gene (Uhlin *et al.*, 1985a, 1985c).

Expression of P-fimbriae is affected by growth conditions such as medium composition and incubation temperature. Glucose repression of P-fimbriae biogenesis appears to involve catabolite repression at the level of *papB* expression, the *papB* gene encoding an activator influencing *papA* gene expression (Uhlin *et al.*, 1985a, 1985c). Transcription of the *papA* gene is regulated in response to growth temperature. That the temperature effect operates at the transcriptional level of subunit biosynthesis rather than at the level of fimbrial assembly was demonstrated using *lacZ* fusion hybrids. It is unclear whether the *papB* gene product is involved in thermoregulation (Göransson and Uhlin, 1984). There are, thus, parallels with the type 1 fimbrial thermoregulation effect.

Regulation in multifimbriate E. coli strains

Clinical isolates of both diarrhoeagenic and uropathogenic *E. coli* have been demonstrated to possess more than one fimbrial antigen, e.g. CS1 or CS2 fimbriae with CS3 fimbriae, K99 and F41 fimbriae, type 1, KS71C and P-fimbriae (Smyth, 1984; Morris *et al.*, 1980; Korhonen *et al.*, 1985a). The presence of more than one fimbrial gene cluster in a strain raises the question of possible interaction between these clusters in the regulation of expression. In addition to strains expressing combinations of adherence antigens, regulation of expression of multiple serotypic variants of a single fimbria needs to be considered (Rhen *et al.*, 1983c). Multiple copies of genes encoding structural subunits or clusters of genes involved in biogenesis may encode the formation of antigenically identical fimbriae, or of several antigenically distinct or cross-reactive fimbriae, or of chimeric fimbriae consisting of a mixture of gene products, all of which have the same receptor specificity (Rhen *et al.*, 1983c, 1985; Rhen, 1985; Hull *et al.*, 1985).

Studies on bacterial populations within a single colony of *E. coli* strain KS71 which had been shown to possess type 1 fimbriae, non-haemagglutinating KS71C fimbriae and two variants of P-fimbriae (Korhonen *et al.*, 1985a; Rhen *et al.*, 1983c; Rhen,

1985), showed that only a minority of the individual cells expressed more than one type of fimbrial antigen (Nowicki *et al.*, 1984). The fact that different fimbriae may apparently not be present on individual cells of a so-called multifimbriate strain, but rather that such a strain may comprise a heterogeneous population of cells expressing different single types or variants of fimbriae, has implications for the regulation of fimbrial phase variation. Separated subpopulations of strain KS71 expressing only one type of fimbriae, obtained using antiserum treatment of a mixed population, gave rise to colonies of bacteria expressing the fimbrial repertoire of the original population (Rhen *et al.*, 1983b, 1983c, 1983d).

The molecular mechanism of this form of fimbrial phase variation *in vitro* is unknown, but the above observations seem to indicate that expression of fimbriae of one antigenic type precludes or represses expression of the others. In the case of the *papA* genes for different antigenic variants or serotypes of P-fimbriae, these may be dispersed as silent genes on the bacterial chromosome. Insertion of a copy of one of these silent *papA* genes into an expression locus or loci may then occur as envisaged for regulation of fimbrial expression in *N. gonorrhoeae* (So *et al.*, 1985a, 1985b). Another possibility is that regulation of phase variation involving site-specific inversion of a controlling element in a manner similar to that described for flagellar phase variation in *Salmonella* may account for the fimbriate to non-fimbriate change or expression of a second *papA* gene encoding the subunit for a different serotype of the fimbriae (Silverman *et al.*, 1979; Silverman and Simon, 1980; Kutsukake and Iino, 1980; Zieg and Simon, 1980; Scott and Simon, 1982). Another alternative explanation of the fimbriate to non-fimbriate phase variation is that a rearrangement event occurs involving deletion of the structural gene from the expression site or deletion of a switch-regulated inducible regulator which would keep the undeleted locus silent as has been demonstrated for *N. gonorrhoeae* (So *et al.*, 1985a 1985b).

In the case of multifimbriate diarrhoeagenic *E. coli*, it has been established in some cases that individual cells express more than one type of fimbria, e.g. CS1 and CS3 fimbriae (Levine *et al.*, 1984), as morphologically and/or serologically distinguishable cell surface structures by electron microscopy. Whether this applies in other instances, e.g. K99 and F41 fimbriae on *E. coli* of O-serovars 9 and 101, requires to be established using electron microscopy with immunolabelling techniques employing monospecific antibodies.

Structural and functional aspects of fimbriae

Fimbriae as receptor-specific adhesins

One area of investigation of *E. coli* fimbriae that has borne fruit in the past 10 years is the identification of the receptors for various types of fimbriae and elucidation of their chemical structures (*Table 3*). Because many rod-shaped, rigid fimbriae on *E. coli* are morphologically indistinguishable yet are composed of subunits of different molecular weight, it has generally been assumed that the subunits possess the specific binding properties of fimbriae or that their assembly into a supramolecular structure confers ligand specificity. Either hypothesis would appear on first principles to satisfactorily explain the binding properties of antigenically different fimbriae with different receptors and also provides an explanation of how type 1 fimbriae or P-fimbriae, for

Table 3. Receptor specificities of various fimbriae of *E. coli*.

Fimbria	*Receptor structure implicated*	*Reference*
Type 1	α-D-Mannosides	Old (1972)
K88	Galactoside	Sellwood (1980)
CFA/I	Sialic acid	Faris *et al.* (1980)
K99	Sialic acid + N-acetylgalactosamine	Lindahl and Wadström (1984); Lindahl *et al.* (1985)
	Neu5Gc-α(2-3)-Gal*p*-β(1-4)-Glc*p*-β(1-1)-ceramide	Smit *et al.* (1984)
P	α-D-Gal(1-4)β-galactoside	Källenius *et al.* (1980b); Svenson *et al.* (1983)
S	Neuraminylα(2-3)galactosides	Parkkinen *et al.* (1983); Korhonen *et al.* (1984)
M	Glycophorin A(N-terminal glyco-octapeptide with essential serine residue and alkali-labile oligosaccharides)	Väisänen *et al.* (1982); Jokinen *et al.* (1985)
987P	Glycoprotein from rabbit small intestinal brush border epithelium; D-galactose and L-fucose	Dean and Isaacson (1985a, 1985b)
Fib	Fibronectin	Fröman *et al.* (1984)

example, can vary antigenically yet retain the same ligand specificity. If the epitope involved in binding is associated with an invariable region of the subunit, ligand specificity might be preserved. Such a proposal has been presented to account for the retention of adhesive specificity by gonococcal fimbriae (Schoolnik *et al.*, 1984; Rothbard *et al.*, 1984). Another theoretical possibility, which might account for retention of the adhesive specificity of a fimbria, is that minor fimbrial components and not the major subunit polypeptide may specify binding properties.

Evidence that gene products other than the *papA* subunit are required for the digalactoside-binding property of P-fimbriae has been provided by the studies of a Swedish group using transposon mutagenesis and frameshift mutagenesis by insertion of a DNA linker (Normark *et al.*, 1983; Norgren *et al.*, 1984; Lindberg *et al.*, 1984; Uhlin *et al.*, 1985a, 1985b, 1985c). Digalactoside-binding properties were retained by *papA* mutants which did not express fimbriae as judged by electron microscopy. Three cistrons *papE*, *papF* and *papG* were identified with digalactoside-specific binding properties. Purified fimbriae from *papF* and *papG* mutants did not bind to digalactoside.

Although formation of fimbriae and specific adhesive properties have been distinguished genetically, direct analysis of purified fimbriae by SDS−PAGE has not provided evidence for the presence of polypeptides other than the *papA* gene product (O'Hanley *et al.*, 1983; Korhonen *et al.*, 1985a). However, by radioiodination of P-fimbriae, Lindberg *et al.* (1985) have detected the presence of the *papE* protein; also, in antiserum to highly purified wild-type P-fimbriae, antibodies to the *papE* and *papF* gene products could be detected.

Using deletion mutants of recombinant plasmids encoding the KS71B type of P-fimbriae, Rhen *et al.* (1986) were independently able to separate fimbriation from haemagglutination of human erythrocytes (P-antigen recognition). The formation of adhesin (haemagglutinin) and fimbriae have also recently been separated in the case of S fimbriae (Hacker *et al.*, 1985). Whether this is a feature common to chromosomally encoded fimbriae remains to be seen. In addition, the 17.6-kd minor component of the K99 fimbria may play some role in its adhesive properties (Mooi and de Graaf, 1985). Advances in this area will undoubtedly come from sequencing data.

Supramolecular structure in relation to antigen variation and immunogenic potential

The fimbriae of *E. coli* can be broadly grouped on a morphological basis: (a) rod-like, rigid filaments of about 6–7 nm in diameter, e.g. type 1, CFA/I, CS1, CS2, 987P fimbriae and (b) delicate, thin, wiry, flexible filaments of about 2–3 nm in diameter, e.g. F41, K88, CS3 fimbriae (Levine *et al.*, 1984). Some adhesins of the latter category have not been easily recognized, e.g. CS3 fimbriae (Smyth, 1982, 1984; Mullany *et al.*, 1983; Levine *et al.*, 1984). Indeed, adhesive protein capsules, termed Z antigens or glycocalyces (capsules?) or afimbrial haemagglutinins, which have been described on some enteropathogenic and uropathogenic strains of *E. coli*, have been shown to comprise meshes of very fine fibrils not unlike those of K88 fimbriae or CS3 fimbriae (Ørskov *et al.*, 1985; Knutton *et al.*, 1984c; Williams *et al.*, 1984; Labigne-Roussel *et al.*, 1984; Freer, 1985) and might be properly considered to be further examples of this fimbrial morphological type.

The supramolecular structure of a type 1 fimbria, as determined from electron microscopy, crystallography and X-ray diffraction, is a right-handed helical array of subunits of molecular weight 17 000 with 3.125 subunits per turn of the helix and a pitch distance of 2.3 nm (Brinton, 1965, 1967). The outer diameter of this fimbria is 7 nm with an axial hole of 2.0–2.5 nm. Other rigid, rod-like fimbriae have the same overall morphological appearance as type 1 fimbriae and are assumed to comprise helical arrays of subunits, albeit that the number of subunits per turn and the pitch distance may vary because of the differences in molecular weights of subunits (Gaastra and de Graaf, 1982; Wilson, 1984; Mooi and de Graaf, 1985; Parry and Rooke, 1985).

The quaternary (supramolecular) structure of type 1 fimbriae is highly resistant to disruption, but it can be dissociated into subunits by boiling under acid conditions and treatment with saturated guanidine hydrochloride (McMichael and Ou, 1979; Eshdat *et al.*, 1981; Dodd and Eisenstein, 1982). Similar difficulties have not been experienced in disaggregating 987P, CFA/I, CS1, CS2 and P-fimbriae, suggesting that the supramolecular structures of these rod-like fimbriae may be different in some respects to that of type 1 fimbriae. The quaternary structures of the thin flexible fimbriae have not been investigated, but seem unlikely to mimic that of type 1 fimbriae from considerations of morphology, subunit molecular weights and restrictions imposed by their diameters. The availability of many fimbriae in highly purified form should allow crystallographic and X-ray diffraction studies to be performed to investigate supramolecular aspects of the structure of other fimbriae.

Monoclonal antibodies raised to type 1 fimbriae have been used to investigate structural and functional properties (Abraham *et al.*, 1983). One hybridoma clone produced monoclonal antibodies that (i) prevented adhesion of the immunizing bacteria to epithelial cells and guinea-pig erythrocytes, (ii) precipitated isolated fimbriae in double diffusion in agar gel, (iii) did not react with fimbrial subunits in immunoelectroblotting and an enzyme-linked immunosorbent assay (ELISA) but did react with reassembled subunits in the ELISA test, and (iv) bound in a highly discrete and periodic manner along the length of intact fimbriae as monitored by immune electron microscopy. Moreover, the binding periodicity pattern of the quaternary structure-specific, anti-adhesive monoclonal antibodies was consistent with the helical structural model for type 1 fimbriae of Brinton (1965, 1967) and the observations of Sweeney and Freer (1979) on the lateral occurrence of binding sites on type 1 fimbriae. In contrast, two other monoclonal antibodies only recognized antigenic epitopes in dissociated fimbrial subunits and did not react with intact fimbriae. These studies demonstrate that antibodies elicited against fimbrial monomers or synthetic polypeptides representing partial primary sequences of subunits may not recognize the quaternary structure of fimbriae as expressed on the cell surface.

Furthermore, Karch *et al.* (1985b) have addressed the issue of antigenic heterogeneity among fimbriae and the relationship between fimbrial structure and the specificity of the immune response with polyclonal antisera raised in rabbits to purified intact fimbriae from several *E. coli* strains. The major findings were that polyclonal antibodies elicited against fimbrial subunits did not recognize intact fimbriae, and that subunits were only poorly recognized by polyclonal anti-fimbrial antisera. Indeed, the reported specificity of antisera raised to intact fimbriae which appear related on the basis of N-terminal or complete amino sequence data, e.g. type 1 and F7 fimbriae (Ørskov *et al.*, 1980b; Ørskov and Ørskov, 1983; Klemm, 1985), seems to indicate that even minor differences in the primary structure of subunits can lead to major differences in the quaternary structure and consequently unique antigenic epitopes expressed on the surface of intact fimbriae. Accordingly, although regions of primary sequences of the K99 fimbrial subunits have been pinpointed by authors to be associated with immunodominant and immunospecific epitopes, these epitopes may be irrelevant in intact fimbriae for protective immunity (Klemm, 1981; Klemm and Mikkelsen, 1982; Gaastra *et al.*, 1983; Dykes *et al.*, 1985). Despite the fact that Worobec *et al.* (1983a) showed that antiserum raised against a dodecapeptide constructed on the basis of the amino acid sequence of a fimbrial subunit reacted with the intact corresponding fimbriae, Karch *et al.* (1985b) caution that the coupling of synthetic polypeptides to a globular carrier could influence the conformation of the peptide and thereby mimic quaternary structural features.

Fimbriae have attracted much attention as potential immunogens because of the realization that fimbriae-mediated adhesion is a primary event for many bacterial infections (Brinton *et al.*, 1978; Gaastra and de Graaf, 1982; Klemm, 1985; Korhonen *et al.*, 1985a). The above-mentioned observations emphasize the importance of the physical structure of fimbriae in their recognition by the host's immune system, if vaccination with purified fimbrial preparations is to be efficacious. The limitations and usefulness of predictions of protein antigenic determinants from amino acid sequences by the method of Hopp and Woods (1981) when applied to fimbrial subunits should be more carefully considered, given the evidence that antibodies to epitopes related to the quaternary structure of fimbriae appear to be important in conferring protection.

Evolutionary relationships between fimbriae

Comparisons of the amino acid sequences of the subunits of purified fimbriae indicate considerable conservation of sequence between those of type 1, $F7_2$, Pap and K99 fimbriae (Klemm, 1985). In contrast, the primary structure of the CFA/I subunit did not show any convincing homology with the sequences for the former fimbrial subunits (Klemm, 1982, 1985). Interestingly, the N-terminal sequence of the subunit of CS2 fimbriae showed extensive homology with that of the CFA/I fimbrial subunit, despite the fact that these two fimbriae showed no immunological cross-reactivity (Klemm *et al.*, 1985b; Smyth, 1982; Cravioto *et al.*, 1982) and their overall amino acid compositions were substantially different (Klemm *et al.*, 1985a). Taken together, the observations suggest caution in interpreting such data, especially if the ultimate aim is to link conserved sequences with a particular function, possibly with a view to the development of synthetic vaccines directed against a number of fimbriae (Klemm, 1985; Mooi and de Graaf, 1985; Worobec *et al.*, 1983b).

Concluding remarks

Molecular cloning techniques, *in vitro* mutagenesis procedures, *in vitro* analysis of gene products, DNA sequencing and probing with monoclonal antibodies have enabled considerable strides to be made in the last few years in our understanding of the antigenicity and functioning of *E. coli* fimbriae as adhesins. The area of molecular pathogenesis has blossomed as a consequence. Knowledge of the process of fimbrial biogenesis and its regulation has increased.

Antigenic variation in *E. coli* fimbriae is close to being explained in terms of the primary, secondary and tertiary structures of fimbrial subunits and in supramolecular terms. It is already clear that immunization with fimbrial antigens can lead in time to the selection of spontaneous antigenic variants which then dominate in an animal population, e.g. K88ad fimbriae. Moreover, phase variation in fimbriae, the genetic mechanism of which in *E. coli* remains to be established, may be important in certain types of *E. coli* infection, e.g. chronic urinary tract infections, where variation in the expression of the adhesins may be related to the occurrence of receptors on host tissues. Also antigenic variation by phase variation may play a role in bacterial virulence by allowing *E. coli* to escape host defences, e.g. phase variation in P-fimbriae, binding of Tamm–Horsfall glycoprotein to type 1 fimbriae.

The production simultaneously of multiple antigenic types of fimbriae with a single ligand specificity, as demonstrated within one colony *in vitro*, may not occur *in vivo*. Rather, antigenic variation may occur *in vivo* by orchestration of gene expression such that the bacterium is one jump ahead of the immunological defence system of the host, in that a subpopulation is expressing antigenically different fimbriae whilst the host is trying to mount an immune response to those of the majority of the population, in a scenario somewhat analogous to antigenic variation in trypanosomes (Spriggs, 1985; Turner, Chapter 1). Indeed, the frequency with which phase variation in bacterial fimbriae occurs *in vitro* is not unlike that with which antigenic variation arises in the variable surface coat glycoproteins of trypanosomes. Minor protein components of fimbriae which act as adhesive components may be immunorecessive *in vivo* under such conditions. As yet, the extent of the antigenic repertoire of chromosomally encoded fimbriae of urinary tract infection *E. coli* is unknown.

Molecular cloning has allowed detailed investigation of the organization of genes involved in the biogenesis of fimbriae and of the primary sequences of subunit polypeptides by DNA sequencing of structural genes. This information should allow understanding of the ways in which fimbrial variation has evolved. Despite all the advances *in vitro*, how expression of fimbrial determinants is regulated under the conditions pertaining *in vivo* at the site of infection remains a challenge to understanding the pathogenesis of bacterial infections caused by *E. coli*.

Acknowledgements

Gillian Johnston is thanked for excellent secretarial service. The author's work is supported by grants from the Medical Research Council of Ireland and a grant from Coralab Research, Cambridge.

References

Abraham,S.N., Hasty,D.L., Simpson,W.A. and Beachey,E.H. (1983) Antiadhesive properties of a quaternary structure-specific hybridoma antibody against type 1 fimbriae of *Escherichia coli*. *J. Exp. Med.*, **158**, 1114–1128.

Abraham,S.N., Babu,J.P., Giampapa,C.S., Hasty,D.L., Simpson,W.A. and Beachey,E.H. (1985) Protection against *Escherichia coli*-induced urinary tract infections with hybridoma antibodies directed against type 1 fimbriae or complementary D-mannose receptors. *Infect. Immun.*, **48**, 625–628.

Bettelheim,K.A. (1978) The sources of 'OH' serotypes of *Escherichia coli*. *J. Hyg.*, **80**, 83–113.

Bijlsma,I.G.W., de Nijs,A. and Frik,J.F. (1981) Adherence of *Escherichia coli* to porcine intestinal brush borders by means of serological variants of the K88 antigen. *Antonie van Leeuwenhoek*, **47**, 467–468.

Bijlsma,I.G.W., de Nijs,A., van der Meer,C. and Frik,J.F. (1982) Different pig phenotypes affect adherence of *Escherichia coli* to jejunal brush borders by K88ab, K88ac, or K88ad antigen. *Infect. Immun.*, **37**, 891–894.

Boylan,M. and Smyth,C.J. (1985) Mobilization of CS fimbriae-associated plasmids of enterotoxigenic *Escherichia coli* of serotype O6:K15:H16 or H- into various wild-type hosts. *FEMS Microbiol. Lett.*, **29**, 83–89.

Boylan,M., Coleman,D. and Smyth,C.J. (1986) Molecular cloning of the CS3 fimbriae determinant of enterotoxigenic *Escherichia coli* of serotype O6:K15:H16 or H-. In *Molecular Biology of Microbial Pathogenicity: Role of Protein–Carbohydrate Interactions*. Lark,D., Normark,S., Uhlin,B.E. and Wolf-Watz,H. (eds), Academic Press, London, in press.

Brinton,C.C., Jr. (1965) The structure, function, synthesis and genetic control of bacterial pili and a molecular model for DNA and RNA transport in Gram negative bacteria. *Trans. N.Y. Acad. Sci.*, **27**, 1003–1054.

Brinton,C.C., Jr. (1967) Contributions of pili to the specificity of the bacterial surface and a unitary hypothesis of conjugal infectious heredity. In *The Specificity of Cell Surfaces*. Davis,B.D. and Warren,L. (eds), Prentice Hall, Inc., Englewood, Cliffs, New Jersey, pp. 37–70.

Brinton,C.C., Jr. (1978) The piliation phase syndrome and the uses of purified pili in disease control. In *XIIIth US–Japan Conference on Cholera*. Miller,C. (ed.), United States Department of Health, Education and Welfare Publication No. NIH 78-1590, Bethesda, MD, pp. 34–60.

Brinton,C.C., Jr. and Thirteen Coauthors. (1978) Uses of pili in gonorrhea control: role of bacterial pili in disease, purification and properties of gonococcal pili, and progress in the development of a gonococcal pilus vaccine for gonorrhea. In *Immunobiology of Neisseria gonorrhoeae*. Brooks,G.F., Gotschlich,E.C., Holmes,K.K., Sawyer,W.D. and Young,F.E. (eds), American Society for Microbiology, Washington, DC, pp. 155–178.

Contrepois,M., Girardeau,J.P., Gouet,P. and Der Vartanian,M. (1983) Expression of K99 pilus of *E. coli*. *Ann. Rech. Vét.*, **14**, 400–407.

Cravioto,A., Scotland,S.M. and Rowe,B. (1982) Hemagglutination activity and colonization factor antigens I and II in enterotoxigenic and non-enterotoxigenic strains of *Escherichia coli* isolated from humans. *Infect. Immun.*, **36**, 189–197.

Darfeuille,A., Lafenille,B., Joly,B. and Cluzel,R. (1983) A new colonization factor antigen (CFA/III) produced by enteropathogenic *Escherichia coli* O128:H12. *Ann. Microbiol. (Inst. Pasteur)*, **134A**, 56–64.

Dean,E.A. and Isaacson,R.E. (1985a) Purification and characterization of a receptor for the 987P pilus of *Escherichia coli*. *Infect. Immun.*, **47**, 98–105.

Dean,E.A. and Isaacson,R.E. (1985b) Location and distribution of a receptor for the 987P pilus of *Escherichia coli* in small intestines. *Infect. Immun.*, **47**, 345–348.
de Graaf,F.K. and Roorda,I. (1982) Production, purification and characterization of the fimbrial adhesive antigen F41 isolated from calf enteropathogenic *Escherichia coli* strain B41M. *Infect. Immun.*, **36**, 751–758.
de Graaf,F.K., Klaasen-Boor,P. and Hees,J.E. (1980) Biosynthesis of the K99 surface antigen is repressed by alanine. *Infect. Immun.*, **30**, 125–128.
de Ree,J.M., Schwillens,P. and van den Bosch,J.F. (1985a) Monoclonal antibodies that recognise the P fimbriae $F7_1$, $F7_2$, F9 and F11 from uropathogenic *Escherichia coli*. *Infect. Immun.*, **50**, 900–904.
de Ree,J.M., Savelkoul,P.H.M., Schwillens,P. and van den Bosch,J.F. (1985b) Monoclonal antibodies raised against five different P fimbriae and type 1A and 1C fimbriae. In *Molecular Biology of Microbial Pathogenicity: Role of Protein–Carbohydrate Interactions*. FEMS Conference Abstracts, Luleå, Sweden, p. P11.
Dodd,D.C. and Eisenstein,B.I. (1982) Antigenic quantitation of type 1 fimbriae on the surface of *Escherichia coli* cells by an enzyme-linked immunosorbent inhibition assay. *Infect. Immun.*, **38**, 764–773.
Duguid,J.P. (1985) Antigens of type-1 fimbriae. In *Immunology of the Bacterial Cell Envelope*. Stewart-Tull,D.E.S. and Davies,M. (eds), John Wiley and Sons, Chichester, pp. 301–318.
Duguid,J.P. and Campbell,I. (1969) Antigens of the type-1 fimbriae of salmonellae and other enterobacteria. *J. Med. Microbiol.*, **2**, 535–553.
Duguid,J.P. and Old,D.C. (1980) Adhesive properties of *Enterobacteriaceae*. In *Bacterial Adherence. Receptors and Recognition Series B*. Beachey,E.H. (ed.), Chapman and Hall, London, Vol. **6**, pp. 185–217.
Duguid,J.P., Smith,I.W., Dempster,G. and Edmunds,P.N. (1955) Non-flagellar filamentous appendages ('fimbriae') and haemagglutinating activity in *Bacterium coli*. *J. Pathol. Bacteriol.*, **70**, 335–348.
Duguid,J.P., Clegg,S. and Wilson,M.I. (1979) The fimbrial and non-fimbrial haemagglutinins of *Escherichia coli*. *J. Med. Microbiol.*, **12**, 213–227.
Dykes,C.W., Halliday,I.J., Read,M.J., Hobden,A.N. and Harford,S. (1985) Nucleotide sequences of four variants of the K88 gene of porcine enterotoxigenic *Escherichia coli*. *Infect. Immun.*, **50**, 279–283.
Echeverria,P., Seriwatana,J., Taylor,D.N., Changchawalit,S., Smyth,C.J., Twohig,J. and Rowe,B. (1986) Plasmids coding for colonization factor antigens I and II, LT, and ST-A2 in *Escherichia coli*. *Infect. Immun.*, **51**, 626– 630.
Eisenstein,B.I. (1981) Phase variation of type 1 fimbriae in *Escherichia coli* is under transcriptional control. *Science*, **214**, 337–339.
Eisenstein,B.I. and Dodd,D.C. (1982) Pseudocatabolite repression of type 1 fimbriae of *Escherichia coli*. *J. Bacteriol.*, **151**, 1560–1567.
Eshdat,Y., Silverblatt,F.J. and Sharon,N. (1981) Dissociation and reassembly of *Escherichia coli* type 1 pili. *J. Bacteriol.*, **148**, 308–314.
Evans,D.G. and Evans,D.J., Jr. (1978) New surface-associated heat-labile colonization factor (CFA/II) produced by enterotoxigenic *Escherichia coli* of serogroups O6 and O8. *Infect. Immun.*, **21**, 638–647.
Evans,D.G., Silver,R.P., Evans,D.J., Jr., Chase,D.G. and Gorbach,S.L. (1975) Plasmid-controlled colonization factor associated with virulence in *Escherichia coli* enterotoxigenic for humans. *Infect. Immun.*, **12**, 656–667.
Evans,D.G., Peñaranda,M.E., Selvidge,L. and Evans,D.J., Jr. (1981) A novel effect of anti-CFA antibody on CFA-positive enterotoxigenic *Escherichia coli*: selective growth of CFA-negative variants. In *Acute Enteric Infections in Children*. Holme,T., Holmgren,J., Merson,M.H. and Möllby,R. (eds), Elsevier, Amsterdam, pp. 96–101.
Fairbrother,J.M., Larivière,S. and Lallier,R. (1986) New fimbrial antigen F165 from *Escherichia coli* serogroup O115 strains isolated from piglets with diarrhea. *Infect. Immun.*, **51**, 10–15.
Faris,A. (1985) Adhesive and hydrophobic adsorptive properties of enterotoxigenic and bovine mastitis *Escherichia coli*: identification of fibronectin-binding fimbriae (Fib fimbriae). Ph.D. Thesis, Swedish University of Agricultural Sciences, Uppsala, Sweden.
Faris,A., Lindahl,M. and Wadström,T. (1980) Gm_2-like glycoconjugates as possible erythrocyte receptors of the CFA/I and K99 haemagglutinins of enterotoxigenic *Escherichia coli*. *FEMS Microbiol. Lett.*, **7**, 265–269.
Faris,A., Sellei,J., Lindahl,M. and Wadström,T. (1982) Haemagglutination of bovine erythrocytes by enterotoxigenic *E. coli* (ETEC) of O6 serogroup: evidence for glycoconjugate receptor heterogeneity. *Zentralbl. Bakteriol. Mikrobiol. Hyg., I Abt. A*, **253**, 175–182.
Freer,J.H. (1985) Illustrated guide to the anatomy of the bacterial cell envelope. In *Immunology of the Bacterial Cell Envelope*. Stewart-Tull,D.E.S. and Davies,M. (eds), John Wiley and Sons, Chichester, pp. 355–383.

Fröman,G., Świtalski,L.M., Faris,A., Wadström,T. and Höök,M. (1984) Binding of *Escherichia coli* to fibronectin: a mechanism of tissue adherence. *J. Biol. Chem.*, **259**, 14899–14905.

Gaastra,W. and de Graaf,F.K. (1982) Host-specific fimbrial adhesins of non-invasive enterotoxigenic *Escherichia coli* strains. *Microbiol. Rev.*, **46**, 129–161.

Gaastra,W., Kleem,P., Walker,J.M. and de Graaf,F.K. (1979) K99 fimbrial proteins: amino- and carboxyl terminal sequences of intact proteins and cyanogen bromide fragments. *FEMS Microbiol. Lett.*, **6**, 15–18.

Gaastra,W., Klemm,P. and de Graaf,F.K. (1983) The nucleotide sequence of the K88ad protein subunit of porcine enterotoxigenic *Escherichia coli*. *FEMS Microbiol. Lett.*, **18**, 177–183.

Gaastra,W., Mooi,F.R., Stuitje,A.R. and de Graaf,F.K. (1981) The nucleotide sequence of the gene encoding the K88ab protein subunit of porcine enterotoxigenic *Escherichia coli*. *FEMS Microbiol. Lett.*, **12**, 41–46.

Gillies,R.R. and Duguid,J.P. (1958) The fimbrial antigens of *Shigella flexneri*. *J. Hyg.*, **56**, 303–318.

Göransson,M. and Uhlin,B.E. (1984) Environmental temperature regulates transcription of a virulence pili operon in *E. coli*. *EMBO J.*, **3**, 2885–2888.

Gross,R.J., Cravioto,A., Scotland,S.M., Cheasty,T. and Rowe,B. (1978) The occurrence of colonisation factor (CF) in enterotoxigenic *Escherichia coli*. *FEMS Microbiol. Lett.*, **3**, 231–233.

Guinée,P.A.M. and Jansen,W.H. (1979) Behaviour of *Escherichia coli* K antigens K88ab, K88ac and K88ad in immunoelectrophoresis, double diffusion, and hemagglutination. *Infect. Immun.*, **23**, 700–705.

Guinée,P.A.M., Veltkamp,J. and Jansen,W.H. (1977) Improved Minca medium for the detection of K99 antigen in calf enterotoxigenic strains of *Escherichia coli*. *Infect. Immun.*, **15**, 676–678.

Guth,B.E.C., Silva,M.L.M., Scaletsky,I.C.A., Toledo,M.R.F. and Trabulsi,L.R. (1985) Enterotoxin production, presence of colonization factor antigen I, and adherence to HeLa cells by *Escherichia coli* O128 strains belonging to different O subgroups. *Infect. Immun.*, **47**, 338–340.

Hacker,J., Schmidt,G., Hughes,C., Knapp,S., Marget,M. and Goebel,W. (1985) Cloning and characterization of genes involved in production of mannose-resistant, neuraminidase-susceptible (X) fimbriae from a uropathogenic O6:K15:H31 *Escherichia coli* strain. *Infect. Immun.*, **47**, 434–440.

Hanley,J., Salit,I.E. and Hofmann,T. (1985) Immunochemical characterization of P pili from invasive *Escherichia coli*. *Infect. Immun.*, **49**, 581–586.

Honda,T., Khan,M.M.A., Takeda,Y. and Miwatani,T. (1983) Grouping of enterotoxigenic *Escherichia coli* by hydrophobicity and its relation to hemagglutination and enterotoxin productions. *FEMS Microbiol. Lett.*, **17**, 273–276.

Honda,T., Arita,M. and Miwatani,T. (1984) Characterization of new hydrophobic pili of human enterotoxigenic *Escherichia coli*: a possible new colonization factor. *Infect. Immun.*, **43**, 959–965.

Hopp,T.P. and Woods,K.R. (1981) Prediction of protein antigenic determinants from amino acid sequences. *Proc. Natl. Acad. Sci. USA*, **78**, 3824–3828.

Houslay,M.D. and Stanley,K.K. (1982) Lipid-protein interactions. In *Dynamics of Biological Membranes*. John Wiley and Sons, Chichester, pp. 92–151.

Hull,R.A., Gill,R.E., Hsu,P., Minshew,B.H. and Falkow,S. (1981) Construction and expression of recombinant plasmids encoding type 1 or D-mannose-resistant pili from a urinary tract infection *Escherichia coli* isolate. *Infect. Immun.*, **33**, 933–938.

Hull,S., Clegg,S., Svanborg-Edén,C. and Hull,R. (1985) Multiple forms of genes in pyelonephritogenic *Escherichia coli* encoding adhesins binding globoseries glycolipid receptors. *Infect. Immun.*, **47**, 80–83.

Isaacson,R.E. (1980) Factors affecting expression of the *Escherichia coli* pilus K99. *Infect. Immun.*, **28**, 190–194.

Isaacson,R.E. (1983) Regulation of expression of *Escherichia coli* pilus K99. *Infect. Immun.*, **40**, 633–639.

Jann,K., Jann,B. and Schmidt,G. (1981) SDS polyacrylamide gel electrophoresis and serological analysis of pili from *Escherichia coli* of different pathogenic origin. *FEMS Microbiol. Lett.*, **11**, 21–25.

Jayappa,H.G., Goodnow,R.A. and Geary,S.J. (1985) Role of *Escherichia coli* type 1 pilus in colonization of porcine ileum and its protective nature as a vaccine antigen in controlling colibacillosis. *Infect. Immun.*, **48**, 350–354.

Jokinen,M., Ehnholm,C., Väisänen-Rhen,V., Korhonen,T.K., Pipkorn,R., Kalkkinen,N. and Gahonberg,C.G. (1985) Identification of the major human red cell sialoglycoprotein, glycophorin A^M, as the receptor for *Escherichia coli* 1H11165 and characterization of the receptor site. *Eur. J. Biochem.*, **147**, 47–52.

Jones,G.W. and Isaacson,R.E. (1984) Proteinaceous bacterial adhesins and their receptors. *CRC Crit. Rev. Microbiol.*, **10**, 229–260.

Källenius,G. and Möllby,R. (1979) Adhesion of *Escherichia coli* to human periurethral cells correlated to mannose-resistant agglutination of human erythrocytes. *FEMS Microbiol. Lett.*, **5**, 295–299.

Källenius,G., Möllby,R. and Winberg,J. (1980a) *In vitro* adhesion of uropathogenic *Escherichia coli* to human periurethral cells. *Infect. Immun.*, **28**, 972–980.
Källenius,G., Möllby,R., Svenson,S.B., Winberg,J., Lundblad,A. and Svenson,S. (1980b) The P^k antigen as receptor of pyelonephritogenic *Escherichia coli. FEMS Microbiol. Lett.*, **7**, 297–302.
Källenius,G., Möllby,R., Svenson,S.B., Helin,I., Hultberg,H., Cedergren,B. and Winberg,J. (1981) Occurrence of P-fimbriated *Escherichia coli* in urinary tract infections. *Lancet*, **II**, 1369–1372.
Karch,H., Leying,H., Büscher,K.-H., Kroll,H.-P. and Opferkuch,W. (1985a) Isolation and separation of physicochemically distinct fimbrial types expressed on a single culture of *Escherichia coli* O7:K1:H6. *Infect. Immun.*, **47**, 549–554.
Karch,H., Leying,H., Goroncy-Bermes,P., Kroll,H.-P. and Opferkuch,W. (1985b) Three-dimensional structure of fimbriae determines specificity of immune response. *Infect. Immun.*, **50**, 517–522.
Kétyi,I. (1984) Non-toxin virulence factors of bacterial enteric pathogens (a review). *Acta Microbiol. Acad. Sci. Hung.*, **31**, 1–25.
Klemm,P. (1981) The complete amino-acid sequence of the K88 antigen, a fimbrial protein from *Escherichia coli. Eur. J. Biochem.*, **117**, 617–627.
Klemm,P. (1982) Primary structure of the CFA1 fimbrial protein from human enterotoxigenic *Escherichia coli* strains. *Eur. J. Biochem.*, **124**, 339–348.
Klemm,P. (1984) The *fimA* gene encoding the type-1 fimbrial subunit of *Escherichia coli*: nucleotide sequence and primary structure of the protein. *Eur. J. Biochem.*, **143**, 395–399.
Klemm,P. (1985) Fimbrial adhesins of *Escherichia coli. Rev. Infect. Dis.*, **7**, 321–340.
Klemm,P. and Mikkelsen,L. (1982) Prediction of antigenic determinants and secondary structures of the K88 and CFA1 fimbrial proteins from enteropathogenic *Escherichia coli. Infect. Immun.*, **38**, 41–45.
Klemm,P., Ørskov,I. and Ørskov,F. (1982) F7 and type 1-like fimbriae from *Escherichia coli* strains isolated from urinary tract infections: protein chemical and immunological aspects. *Infect. Immun.*, **36**, 462–468.
Klemm,P., Ørskov,I. and Ørskov,F. (1983) Isolation and characterization of the F12 adhesive fimbrial antigen from uropathogenic *Escherichia coli* strains. *Infect. Immun.*, **40**, 91–96.
Klemm,P., Gaastra,W., McConnell,M.M. and Smith,H.R. (1985a) The CS2 fimbrial antigen from *Escherichia coli*, purification, characterization and partial covalent structure. *FEMS Microbiol. Lett.*, **26**, 207–210.
Klemm,P., Jørgensen,B.J., van Die,I., de Ree,H. and Bergmans,H. (1985b) The *fim* genes responsible for synthesis of type 1 fimbriae in *Escherichia coli*, cloning and genetic organization. *Mol. Gen. Genet.*, **199**, 410–414.
Knutton,S., Lloyd,D.R., Candy,D.C.A. and McNeish,A.S. (1984a) *In vitro* adhesion of enterotoxigenic *Escherichia coli* to human intestinal epithelial cells from mucosal biopsies. *Infect. Immun.*, **44**, 514–518.
Knutton,S., Lloyd,D.R., Candy,D.C.A. and McNeish,A.S. (1984b) Ultrastructural study of adhesion of enterotoxigenic *Escherichia coli* to erythrocytes and human intestinal epithelial cells. *Infect. Immun.*, **44**, 519–527.
Knutton,S., Williams,P.H., Lloyd,D.R., Candy,D.C.A. and McNeish,A.S. (1984c) Ultrastructural study of adherence to and penetration of cultured cells by two invasive *Escherichia coli* strains isolated from infants with enteritis. *Infect. Immun.*, **44**, 599–608.
Knutton,S., Lloyd,D.R., Candy,D.C.A. and McNeish,A.S. (1985) Adhesion of enterotoxigenic *Escherichia coli* to human small intestinal enterocytes. *Infect. Immun.*, **48**, 824–831.
Korhonen,T.K., Nurmiaho,E.-L., Ranta,H. and Svanborg-Edén,C. (1980) New method for isolation of immunologically pure pili from *Escherichia coli. Infect. Immun.*, **27**, 569–575.
Korhonen,T.K., Leffler,H. and Svanborg-Edén,C. (1981) Binding specificity of piliated strains of *Escherichia coli* and *Salmonella typhimurium* to epithelial cells, *Saccharomyces cerevisiae* cells and erythrocytes. *Infect. Immun.*, **32**, 796–804.
Korhonen,T.K., Väisänen,V., Saxén,H., Hultberg,H. and Svenson,S.B. (1982) P-antigen-recognizing fimbriae from human uropathogenic *Escherichia coli* strains. *Infect. Immun.*, **37**, 286–291.
Korhonen,T.K., Väisänen-Rhen,V., Rhen,M., Pere,A., Parkkinen,J. and Finne,J. (1984) *Escherichia coli* fimbriae recognizing sialyl galactosides. *J. Bacteriol.*, **159**, 762–766.
Korhonen,T.K., Rhen,M., Väisänen-Rhen,V. and Pere,A. (1985a) Antigenic and functional properties of enterobacterial fimbriae. In *Immunology of the Bacterial Cell Envelope.* Stewart-Tull,D.E.S. and Davies,M. (eds), John Wiley and Sons, Chichester, pp. 319–354.
Korhonen,T.K., Valtonen,M.V., Parkkinen,J., Väisänen-Rhen,V., Finne,J., Ørskov,F., Ørskov,I., Svenson, S.B. and Mäkelä,P.H. (1985b) Serotypes, hemolysin production, and receptor recognition of *Escherichia coli* strains associated with neonatal sepsis and meningitis. *Infect. Immun.*, **48**, 486–491.
Korhonen,T.K., Nowicki,B., Väisänen-Rhen,V., Rhen,M., Pere,A., Westerlund,B. and Mäkelä,P.H. (1985c) Fimbriae on *Escherichia coli* strains from human extraintestinal infections: receptor specificities, occur-

rence and phase variation. In *Molecular Biology of Microbial Pathogenicity: Role of Protein–Carbohydrate Interactions.* FEMS Conference Abstracts, Luleå, Sweden, p. 34.
Kuriyama,S.M. and Silverblatt,F.J. (1986) Effect of Tamm–Horsfall urinary glycoprotein on phagocytosis and killing of type I-fimbriated *Escherichia coli. Infect. Immun.*, **51**, 193–198.
Kusecek,B., Wloch,H., Mercer,A., Väisänen,V., Pluschke,G., Korhonen,T. and Achtman,M. (1984) Lipopolysaccharide, capsule, and fimbriae as virulence factors among O1, O7, O16, O18 or O75 and K1, K5, or K100 *Escherichia coli. Infect. Immun.*, **43**, 368–379.
Kutsukake,K. and Iino,T. (1980) A *trans*-acting factor mediates inversion of a specific DNA segment in flagellar phase variation of *Salmonella. Nature,* **284**, 479–481.
Labigne-Roussel,A.F., Lark,D., Schoolnik,G. and Falkow,S. (1984) Cloning and expression of an afimbrial adhesin (AFA-I) responsible for P blood group-independent, mannose-resistant hemagglutination from a pyelonephritogenic *Escherichia coli* strain. *Infect. Immun.*, **46**, 251–259.
Leffler,H. and Svanborg-Edén,C. (1980) Chemical identification of a glycosphingolipid receptor for *Escherichia coli* attaching to human urinary tract epithelial cells and agglutinating human erythrocytes. *FEMS Microbiol. Lett.*, **8**, 127–134.
Levine,M.M., Black,R.E., Brinton,C.C., Clements,M.L., Fusco,P., Hughes,T.P., O'Donnell,S., Robins-Browne,R., Mood,S. and Young,C.R. (1982) Reactogenicity, immunogenicity and efficacy studies on *Escherichia coli* type 1 somatic pili parenteral vaccine in man. *Scand. J. Infect. Dis.*, Suppl. **33**, 85–95.
Levine,M.M., Kaper,J.B., Black,R.E. and Clements,M.L. (1983a) New knowledge on pathogenesis of bacterial enteric infections as applied to vaccine development. *Microbiol. Rev.*, **47**, 510–550.
Levine,M.M., Ristaino,P., Sack,R.B., Kaper,J.B., Ørskov,F. and Ørskov,I. (1983b) Colonization factor antigens I and II and type 1 somatic pili in enterotoxigenic *Escherichia coli*: relation to enterotoxin type. *Infect. Immun.*, **39**, 889–897.
Levine,M.M., Ristaino,P., Marley,C., Smyth,C., Knutton,S., Boedeker,E., Black,R., Young,C., Clements, M.L., Cheney,C. and Patnaik,R. (1984) Coli surface antigens 1 and 3 of colonization factor antigen II-positive enterotoxigenic *Escherichia coli*: morphology, purification and immune responses in humans. *Infect. Immun.*, **44**, 409–420.
Lindahl,M. and Wadström,T. (1984) K99 surface haemagglutinin of enterotoxigenic *E. coli* recognize terminal N-acetylgalactosamine and sialic acid residues of glycophorin and other complex glycoconjugates. *Vet. Microbiol.*, **9**, 249–257.
Lindahl,M., Carlstedt,I. and Wadström,T. (1985) Binding of K99 fimbriae to pig small intestinal mucin and erythrocyte membrane glycoproteins. In *Molecular Biology of Microbial Pathogenicity: Role of Protein–Carbohydrate Interactions.* FEMS Conference Abstracts, Luleå, Sweden, p. P25.
Lindberg,F.P., Lund,B. and Normark,S. (1984) Genes of pyelonephritogenic *E. coli* required for digalactoside-specific agglutination of human cells. *EMBO J.*, **3**, 1167–1173.
Lindberg,F., Lund,B. and Normark,S. (1985) Gene products essential for specific adhesion of Pap-pili are pilin-like proteins present in the purified pili. In *Molecular Biology of Microbial Pathogenicity: Role of Protein–Carbohydrate Interactions.* FEMS Conference Abstracts, Luleå, Sweden, p. P20.
Low,D., David,V., Lark,D., Schoolnik,G. and Falkow,S. (1984) Gene clusters governing the production of hemolysin and mannose-resistant hemagglutination are closely linked in *Escherichia coli* serotype O4 and O6 isolates from urinary tract infections. *Infect. Immun.*, **43**, 353–358.
Manning,P.A., Timmis,K.N. and Stevenson,G. (1985) Colonization factor antigen II (CFA/II) of enterotoxigenic *Escherichia coli*: molecular cloning of the CS3 determinant. *Mol. Gen. Genet.*, **200**, 322–327.
McElhaney,R.N. (1982) Effects of membrane lipids on transport and enzymic activities. In *Current Topics in Membranes and Transport.* Razin,S. and Rottem,S. (eds), Academic Press, New York, Vol. **17**, pp. 317–380.
McElhaney,R.N. (1984) The relationship between membrane lipid fluidity and phase state and the ability of bacteria and mycoplasma to grow and survive at various temperatures. In *Membrane Fluidity.* Kates,M. (ed.), Plenum Publishing Corporation, New York, pp. 249–278.
McMichael,J.C. and Ou,J.T. (1979) Structure of common pili from *Escherichia coli. J. Bacteriol.*, **138**, 969–975.
Mooi,F.R. and de Graaf,F.K. (1985) Molecular biology of fimbriae of enterotoxigenic *Escherichia coli. Curr. Top. Microbiol. Immunol.*, **118**, 119–138.
Mooi,F.R., de Graaf,F.K. and van Embden,J.D.A. (1979) Cloning, mapping and expression of the genetic determinant that encodes for the K88ab antigen. *Nucleic Acids Res.*, **6**, 849–865.
Moon,H.W., Kohler,E.M., Schneider,R.A. and Whipp,S.C. (1980) Prevalence of pilus antigens, enterotoxin types and enteropathogenicity among K88-negative enterotoxigenic *Escherichia coli* from neonatal pigs. *Infect. Immun.*, **27**, 222–230.

Morris,J.A., Stevens,A.E. and Sojka,W.J. (1978) Anionic and cationic components of the K99 surface antigen from *Escherichia coli* B41. *J. Gen. Microbiol.*, **107**, 173–175.
Morris,J.A., Thorns,C.J. and Sojka,W.J. (1980) Evidence for two adhesive antigens on the K99 reference strain *Escherichia coli* B41. *J. Gen. Microbiol.*, **118**, 107–113.
Morris,J.A., Thorns,C., Scott,A.C., Sojka,W.J. and Wells,G.A. (1982) Adhesion *in vitro* and *in vivo* associated with an adhesive antigen (F41) produced by a K99 mutant of the reference strain *Escherichia coli* B41. *Infect. Immun.*, **36**, 1146–1153.
Morris,J.A., Thorns,C.J., Wells,G.A.W. and Sojka,W.J. (1983) The production of F41 fimbriae by piglet strains of enterotoxigenic *Escherichia coli* that lack K88, K99 and 987P fimbriae. *J. Gen. Microbiol.*, **129**, 2753–2759.
Mullany,P., Field,A.M., McConnell,M.M., Scotland,S.M., Smith,H.R. and Rowe,B. (1983) Expression of plasmids coding for colonization factor antigen II (CFA/II) and enterotoxin production in *Escherichia coli*. *J. Gen. Microbiol.*, **129**, 3591–3601.
Nagy,B., Moon,H.W. and Isaacson,R.E. (1976) Colonization of porcine small intestine by *Escherichia coli*: ileal colonization and adhesion of pig enteropathogens that lack K99 antigen and by some acapsular mutants. *Infect. Immun.*, **13**, 1214–1220.
Nagy,B., Moon,H.W. and Isaacson,R.E. (1977) Colonization of porcine intestine by enterotoxigenic *Escherichia coli*: selection of piliated forms *in vivo*, adhesion of piliated forms to epithelial cells *in vitro*, and incidence of a pilus antigen among porcine enteropathogenic *E. coli*. *Infect. Immun.*, **16**, 344–352.
Norgren,M., Normark,S., Lark,D., O'Hanley,P., Schoolnik,G., Falkow,S., Svanborg-Edén,C., Båga,M. and Uhlin,B.E. (1984) Mutations in *E. coli* cistrons affecting adhesion to human cells do not abolish Pap pili fiber formation. *EMBO J.*, **3**, 1159–1165.
Normark,S., Lark,D., Hull,R., Norgren,M., Båga,M., O'Hanley,P., Schoolnik,G. and Falkow,S. (1983) Genetics of digalactoside-binding adhesin from a uropathogenic *Escherichia coli* strain. *Infect. Immun.*, **41**, 942–949.
Normark,S., Båga,M., Göransson,M., Lindberg,F., Lund,B., Norgren,M. and Uhlin,B.-E. (1985) Minor pilus components acting as adhesins. In *Molecular Biology of Microbial Pathogenicity: Role of Protein–Carbohydrate Interactions.* FEMS Conference Abstracts, Luleå, Sweden, p. 5.
Nowicki,B., Rhen,M., Väisänen-Rhen,V., Pere,A. and Korhonen,T.K. (1984) Immunofluorescence study of fimbrial phase variation in *Escherichia coli* KS71. *J. Bacteriol.*, **160**, 691–695.
Nowotarska,M. and Mulczyk,M. (1977) Serologic relationship of fimbriae among *Enterobacteriaceae*. *Arch. Immunol. Ther. Exp.*, **25**, 7–16.
Ofek,I., Mosek,A. and Sharon,N. (1981) Mannose-specific adherence of *Escherichia coli* freshly excreted in the urine of patients with urinary infections, and of isolates subcultured from the infected urine. *Infect. Immun.*, **34**, 708–711.
O'Hanley,P., Lark,D., Normark,S., Falkow,S. and Schoolnik,G.K. (1983) Mannose-sensitive and gal-gal binding *Escherichia coli* pili from recombinant strains: chemical, functional and serological properties. *J. Exp. Med.*, **158**, 1713–1719.
Old,D.C. (1972) Inhibition of the interaction between fimbrial haemagglutinins and erythrocytes by D-mannose and other carbohydrates. *J. Gen. Microbiol.*, **71**, 149–157.
Orndorff,P.E. and Falkow,S. (1984) Organization and expression of genes responsible for type 1 piliation in *Escherichia coli*. *J. Bacteriol.*, **159**, 736–744.
Orndorff,P.E., Spears,P.A., Schauer,D. and Falkow,S. (1985) Two modes of control of *pilA*, the gene encoding type 1 pilin in *Escherichia coli*. In *Molecular Biology of Microbial Pathogenicity: Role of Protein–Carbohydrate Interactions.* FEMS Conference Abstracts, Luleå, Sweden, p. 9.
Ørskov,F. and Ørskov,I. (1978) Special *Escherichia coli* serotypes from enteropathies in domestic animals and man. *Zentralbl. Veterinarmed. Reihe A*, **29**, Suppl., 7–14.
Ørskov,I. and Ørskov,F. (1983) Serology of *Escherichia coli* fimbriae. *Prog. Allergy*, **33**, 80–105.
Ørskov,I., Ørskov,F, Sojka,W.J. and Leach,J.N. (1961) Simultaneous occurrence of *E. coli* B and L antigens in strains from diseased swine. Influence of cultivation temperature on two new *E. coli* K antigens K87 and K88. *Acta Pathol. Microbiol. Scand.*, **53**, 404–422.
Ørskov,I., Ørskov,F., Sojka,W.J. and Wittig,W. (1964) K antigens K88ab(L) and K88ac(L) in *E. coli*. A new O antigen: O147 and a new K antigen: K89 (B). *Acta Pathol. Microbiol. Scand.*, **62**, 439–447.
Ørskov,I., Ferencz,A. and Ørskov,F. (1980a) Tamm–Horsfall protein or uromucoid is the normal urinary slime that traps type 1 fimbriated *Escherichia coli*. *Lancet*, **I**, 887.
Ørskov,I., Ørskov,F. and Birch-Andersen,A. (1980b) Comparison of *Escherichia coli* fimbrial antigen F7 with type 1 fimbriae. *Infect. Immun.*, **27**, 657–666.

Ørskov,I., Ørskov,F., Birch-Andersen,A., Klemm,P. and Svanborg-Edén,C. (1982a) Protein attachment factors; fimbriae in adhering *Escherichia coli* strains. In *Bacterial Vaccines. Seminars in Infectious Diseases.* Weinstein,L. and Fields,B.N. (eds), Thieme-Stratton, New York, Vol. **IV**, pp. 97–104.

Ørskov,I., Ørskov,F., Birch-Andersen,A., Kanamori,M. and Svanborg-Edén,C. (1982b) O, K, H and fimbrial antigens in *Escherichia coli* serotypes associated with pyelonephritis and cystitis. *Scand. J. Infect. Dis., Suppl.* **33**, 18–24.

Ørskov,I., Birch-Andersen,A., Duguid,J.P., Stenderup,J. and Ørskov,F. (1985) An adhesive protein capsule of *Escherichia coli. Infect. Immun.*, **47**, 191–200.

Parkkinen,J., Finne,J., Achtman,M., Väisänen,V. and Korhonen,T.K. (1983) *Escherichia coli* strains binding neuraminyl α2-3 galactosides. *Biochem. Biophys. Res. Commun.*, **111**, 456–461.

Parry,S.H. and Porter,P. (1978) Immunological aspects of cell membrane adhesion demonstrated by porcine enteropathogenic *Escherichia coli. Immunology,* **34**, 41–49.

Parry,S.H. and Rooke,D.M. (1985) Adhesins and colonization factors of *Escherichia coli*. In *The Virulence of Escherichia coli: Reviews and Methods.* Sussman,M. (ed.), Special Publications of the Society for General Microbiology No. 13, Academic Press, London, pp. 79–155.

Pearce,W.A. and Buchanan,T.M. (1980) Structure and cell membrane-binding properties of bacterial fimbriae. In *Bacterial Adherence. Receptors and Recognition Series B.* Beachey,E.H. (ed.), Chapman and Hall, London, Vol. **6**, pp. 289–344.

Peñaranda,M.E., Mann,M.B., Evans,D.G. and Evans,D.J., Jr. (1980) Transfer of an ST:LT:CFA/II plasmid into *Escherichia coli* K-12 strain RR1 by cotransformation with pSC301 plasmid DNA. *FEMS Microbiol. Lett.*, **8**, 251–254.

Peñaranda,M.E., Evans,D.G., Murray,B.E. and Evans,D.J., Jr. (1983) ST:LT:CFA/II plasmids in enterotoxigenic *Escherichia coli* belonging to serogroups O6, O8, O85 and O139. *J. Bacteriol.*, **154**, 980–983.

Pere,A., Väisänen-Rhen,V., Rhen,M., Tenhunen,J. and Korhonen,T.K. (1985) P fimbriae on pyelonephritogenic *Escherichia coli* O2, O4 and O6 strains: analysis by immune precipitation. In *Molecular Biology of Microbial Pathogenicity: Role of Protein–Carbohydrate Interactions.* FEMS Conference Abstracts, Luleå, Sweden, p. P13.

Rhen,M. (1985) Characterization of DNA fragments encoding fimbriae of the uropathogenic *Escherichia coli* strain KS71. *J. Gen. Microbiol.*, **131**, 571–580.

Rhen,M., Klemm,P., Wahlström,E., Svenson,S.B., Källenius,G. and Korhonen,T.K. (1983a) P fimbriae of *Escherichia coli*: immuno- and protein chemical characterization of fimbriae from two pyelonephritogenic strains. *FEMS Microbiol. Lett.*, **18**, 233–238.

Rhen,M., Wahlström,E. and Korhonen,T.K. (1983b) P fimbriae of *Escherichia coli*: fractionation by immune precipitation. *FEMS Microbiol. Lett.*, **18**, 227–232.

Rhen,M., Knowles,J., Penttinen,M., Sarvas,M. and Korhonen,T.K. (1983c) P fimbriae of *Escherichia coli*: molecular cloning of DNA fragments containing the structural genes. *FEMS Microbiol. Lett.*, **19**, 119–123.

Rhen,M., Mäkelä,P.H. and Korhonen,T.K. (1983d) P fimbriae of *Escherichia coli* are subject to phase variation. *FEMS Microbiol. Lett.*, **19**, 267–271.

Rhen,M., Väisänen,V., Pere,A. and Korhonen,T.K. (1985) Complementation and regulatory interaction between two cloned fimbrial clusters of *Escherichia coli* strain KS71. *Mol. Gen. Genet.*, **200**, 60–64.

Rhen,M., Tenhunen,J., Väisänen-Rhen,V., Pere,A., Båga,M. and Korhonen,T.K. (1986) Fimbriation and P-antigen recognition of *Escherichia coli* strains harbouring mutated recombinant plasmids encoding fimbrial adhesins of the uropathogenic *E. coli* strain KS71. *J. Gen. Microbiol.*, **132**, 71–77.

Rothbard,J.B., Fernandez,R. and Schoolnik,G.K. (1984) Strain-specific and common epitopes of gonococcal pili. *J. Exp. Med.*, **160**, 208–221.

Saier,M.H., Schmidt,M.R. and Leibowitz,M. (1978) Cyclic AMP-dependent synthesis of fimbriae in *Salmonella typhimurium*: effects of *cys* and *pts* mutations. *J. Bacteriol.*, **134**, 356–358.

Salit,I.E., Vavougios,J. and Hofmann,T. (1983) Isolation and characterization of *Escherichia coli* pili from diverse clinical sources. *Infect. Immun.*, **42**, 755–762.

Schmitz,S., Abe,C., Moser,I., Ørskov,I., Ørskov,F., Jann,B. and Jann,K. (1986) Monoclonal antibodies against the non-hemagglutinating fimbrial antigen 1C (pseudotype 1) of *Escherichia coli. Infect. Immun.*, **51**, 54–59.

Schoolnik,G.K., Fernandez,R., Tai,J.Y., Rothbard,J. and Gotschlich,E.C. (1984) Gonococcal pili: primary structure and receptor binding domain. *J. Exp. Med.*, **159**, 1351–1370.

Scotland,S.M., Gross,R.J. and Rowe,B. (1977) Serotype-related enterotoxigenicity in *Escherichia coli* O6:H16 and O148:H28. *J. Hyg.*, **79**, 395–403.

Scotland,S.M., Day,N.P., Cravioto,A., Thomas,L.V. and Rowe,B. (1981) Production of heat-labile or heat-

stable enterotoxins by strains of *Escherichia coli* belonging to serogroups O44, O114, and O128. *Infect. Immun.*, **31**, 500–503.
Scotland,S.M., McConnell,M.M., Willshaw,G.A., Rowe,B. and Field,A.M. (1985) Properties of wild-type strains of enterotoxigenic *Escherichia coli* which produce colonisation factor antigen II, and belong to serogroups other than O6. *J. Gen. Microbiol.*, **131**, 2327–2333.
Scott,T.N. and Simon,M.I. (1982) Genetic analysis of the mechanism of the *Salmonella* phase variation site specific recombination system. *Mol. Gen. Genet.*, **188**, 313–321.
Sellwood,R. (1980) The interaction of the K88 antigen with porcine intestinal epithelial cell brush-borders. *Biochim. Biophys. Acta*, **632**, 326–335.
Sherman,P.M., Houston,W.C. and Boedeker,E.C. (1985) Functional heterogeneity of intestinal *Escherichia coli* strains expressing type 1 somatic pili (fimbriae): assessment of bacterial adherence to intestinal membranes and surface hydrophobicity. *Infect. Immun.*, **49**, 797–804.
Silverman,M. and Simon,M. (1980) Phase variation: genetic analysis of switching mutants. *Cell*, **19**, 845–854.
Silverman,M., Zieg,J., Hilmen,M. and Simon,M. (1979) Phase variation in *Salmonella*: genetic analysis of a recombinational switch. *Proc. Natl. Acad. Sci. USA*, **76**, 391–395.
Smit,H., Gaastra,W., Kamerling,J.P., Vliegenthart,J.F.G. and de Graaf,F.K. (1984) Isolation and structural characterization of the equine receptor for enterotoxigenic *Escherichia coli* K99 fimbrial adhesin. *Infect. Immun.*, **46**, 578–584.
Smith,H.R., Willshaw,G.A. and Rowe,B. (1982) Mapping of a plasmid, coding for colonization factor antigen I and heat-stable enterotoxin production, isolated from an enterotoxigenic strain of *Escherichia coli*. *J. Bacteriol.*, **149**, 264–275.
Smith,H.R., Scotland,S.M. and Rowe,B. (1983) Plasmids that code for production of colonization factor antigens II and enterotoxin production in strains of *Escherichia coli*. *Infect. Immun.*, **40**, 1236–1239.
Smith,H.R., Scotland,S.M. and Rowe,B. (1985) Genetics of *Escherichia coli* virulence. In *The Virulence of Escherichia coli: Reviews and Methods*. Sussman,M. (ed.), Special Publications of the Society for General Microbiology No. 13, Academic Press, London, pp. 79–155.
Smyth,C.J. (1982) Two mannose-resistant haemagglutinins on enterotoxigenic *Escherichia coli* of serotype O6:K15:H16 or H- isolated from travellers' and infantile diarrhoea. *J. Gen. Microbiol.*, **128**, 2081–2096.
Smyth,C.J. (1984) Serologically distinct fimbriae on enterotoxigenic *Escherichia coli* of serotype O6:K15:H16 or H-. *FEMS Microbiol. Lett.*, **21**, 51–57.
So,M., Heffron,F. and McCarthy,B.J. (1979) The *E. coli* gene encoding heat stable toxin is a bacterial transposon flanked by inverted repeats of IS*1*. *Nature*, **277**, 453–456.
So,M., Billyard,E., Meyer,T.F. and Segal,E. (1985a) Regulation of pilus phase variation in *Neisseria gonorrhoeae*. In *Microbiology — 1985*. Leive,L. (ed.), American Society for Microbiology, Washington, DC, pp. 287–291.
So,M., Billyard,E., Deal,C., Getzoff,E., Hagblom,P., Meyer,T.F., Segal,E. and Tainer,J. (1985b) Gonococcal pilus: genetics and structure. *Curr. Top. Microbiol. Immunol.*, **118**, 13–28.
Spriggs,D.R. (1985) Antigenic variation in trypanosomes: genomes in flux. *J. Infect. Dis.*, **152**, 855–856.
Svanborg-Edén,C. and Hansson,H.A. (1978) *Escherichia coli* pili as possible mediators of attachment to human urinary tract epithelial cells. *Infect. Immun.*, **21**, 229–237.
Svanborg-Edén,C., Eriksson,B. and Hansson,L.Å. (1977) Adhesion of *Escherichia coli* to human uroepithelial cells *in vitro*. *Infect. Immun.*, **18**, 767–774.
Svenson,S.B. and Källenius,G. (1983) Density and localization of P-fimbriae-specific receptors on mammalian cells: fluorescence-activated cell analysis. *Infection*, **11**, 6–12.
Svenson,S.B., Källenius,G., Möllby,R., Hultberg,H. and Winberg,J. (1982) Rapid identification of P-fimbriated *Escherichia coli* by a receptor specific particle agglutination test. *Infection*, **4**, 209–214.
Svenson,S.B., Hultberg,H., Källenius,G., Korhonen,T.K., Möllby,R. and Winberg,J. (1983) P fimbriae of pyelonephritogenic *Escherichia coli*: identification and chemical characterization of receptors. *Infection*, **11**, 61–67.
Sweeney,G. and Freer,J.H. (1979) Location of binding sites on common type 1 fimbriae of *Escherichia coli*. *J. Gen. Microbiol.*, **112**, 321–328.
Thomas,L.V. and Rowe,B. (1982) The occurrence of colonisation factors (CFA/I, CFA/II and E8775) in enterotoxigenic *Escherichia coli* from various countries in South East Asia. *Med. Microbiol. Immunol.*, **171**, 85–90.
Thomas,L.V., Cravioto,A., Scotland,S.M. and Rowe,B. (1982) New fimbrial antigenic type (E8775) that may represent a colonization factor in enterotoxigenic *Escherichia coli* in humans. *Infect. Immun.*, **35**, 1119–1124.

Thomas,L.V., McConnell,M.M., Rowe,B. and Field,A.M. (1985) The possession of three novel coli surface antigens by enterotoxigenic *Escherichia coli* strains positive for the putative colonisation factor PCF8775. *J. Gen. Microbiol.*, **131**, 2319–2326.
To,S.C-M., Moon,H.W. and Runnels,P.L. (1984) Type 1 pili (F1) of porcine enterotoxigenic *Escherichia coli*: vaccine trial and tests for production in the small intestine during disease. *Infect. Immun.*, **43**, 1–5.
Uhlin,B.E., Båga,M., Göransson,M., Lindberg,F.P., Lund,B., Norgren,M. and Normark,S. (1985a) Genes determining adhesin formation in uropathogenic *Escherichia coli*. *Curr. Top. Microbiol. Immunol.*, **118**, 163–178.
Uhlin,B.E., Norgren,M., Båga,M. and Normark,S. (1985b) Adhesion to human cells by *Escherichia coli* lacking the major subunit of a digalactoside-specific pilus adhesin. *Proc. Natl. Acad. Sci. USA*, **82**, 1800–1804.
Uhlin,B.E., Båga,M., Göransson,M., Lindberg,F., Lund,B., Norgren,M. and Normark,S. (1985c) Regulation and biogenesis of digalactoside-binding pili. In *Molecular Biology of Microbial Pathogenicity: Role of Protein–Carbohydrate Interactions*. FEMS Conference Abstracts, Luleå, Sweden, p. 7.
Väisänen-Rhen,V. (1984) Fimbria-like haemagglutinin of *Escherichia coli* O75 strains. *Infect. Immun.*, **46**, 401–407.
Väisänen,V., Elo,J., Tallgren,L.G., Siitonen,A., Mäkelä,P.H., Svanborg-Edén,C., Källenius,G., Svenson, S.B., Hultberg,H. and Korhonen,T.K. (1981) Mannose resistant haemagglutination and P antigen recognition are characteristic of *Escherichia coli* causing primary pyelonephritis. *Lancet*, **II**, 1366–1369.
Väisänen,V., Korhonen,T.K., Jokinen,M., Gahmberg,C.G. and Ehnholm,C. (1982) Blood group M-specific haemagglutinin in pyelonephritogenic *Escherichia coli*. *Lancet*, **I**, 1192.
Väisänen-Rhen,V., Elo,J., Väisänen,E., Siitonen,A., Ørskov,I., Ørskov,F., Svenson,S.B., Mäkelä,P.H. and Korhonen,T.K. (1984) P-fimbriated clones among uropathogenic *Escherichia coli* strains. *Infect. Immun.*, **43**, 149–155.
van Embden,J.D.A., de Graaf,F.K., Schouls,L.M. and Teppema,J.S. (1980) Cloning and expression of a deoxyribonucleic acid fragment that encodes the adhesive antigen K99. *Infect. Immun.*, **29**, 1125–1133.
van Verseveld,H.W., Bakker,P., van der Woude,T., Terleth,C. and de Graaf,F.K. (1985) Production of fimbrial adhesins K99 and F41 by enterotoxigenic *Escherichia coli* as a function of growth-rate domain. *Infect. Immun.*, **49**, 159–163.
Wadström,T. and Trust,T.J. (1984) Bacterial surface lectins. In *Medical Microbiology*. Easmon,C.S.F. and Jeljaszawicz,J. (eds), Academic Press, London, Vol. **4**, pp. 287–334.
Wadström,T., Faris,A., Lindahl,M. and Hjertén,S. (1984) Hydrophobic properties of pili. In *Attachment of Organisms to the Gut Mucosa*. Boedeker,E.C. (ed.), CRC Press, Inc., Boca Raton, pp. 113–120.
Williams,P.H., Knutton,S., Brown,M.G.M., Candy,D.C.A. and McNeish,A.S. (1984) Characterization of nonfimbrial mannose-resistant protein haemagglutinins of two *Escherichia coli* strains isolated from infants with enteritis. *Infect. Immun.*, **44**, 592–598.
Willshaw,G.A., Smith,H.R., McConnell,M.M. and Rowe,B. (1985) Expression of cloned plasmid regions encoding colonization factor antigen I (CFA/I) in *Escherichia coli*. *Plasmid*, **13**, 8–16.
Wilson,R.A. (1984) *Escherichia coli* fimbriae–structure, function and nomenclature. In *Proceedings of the Fourth International Symposium on Neonatal Diarrhea*. Acres,S.D. (ed.) Veterinary Infectious Disease Organization, Saskatoon, Canada, pp. 610–624.
Worobec,E.A., Taneja,A.K., Hodges,R.S. and Paranchych,W. (1983a) Localization of the major antigen determinant of EDP208 pili at the N-terminus of the pilus protein. *J. Bacteriol.*, **153**, 955–961.
Worobec,E.A., Shastry,P., Smart,W., Bradley,R., Singh,B. and Paranchych,W. (1983b) Monoclonal antibodies against colonization factor antigen I pili from enterotoxigenic *Escherichia coli*. *Infect. Immun.*, **41**, 1296–1301.
Zieg,J. and Simon,M. (1980) Analysis of the nucleotide sequence of an invertible controlling element. *Proc. Natl. Acad. Sci. USA*, **77**, 4196–4200.

CHAPTER 8

Molecular basis for antigenic variation in a relapsing fever Borrelia

RONALD H.A.PLASTERK[1], MELVIN I.SIMON[1] and ALAN G.BARBOUR[2]

[1]*Division of Biology, California Institute of Technology, Pasadena, CA 91125, and* [2]*National Institutes of Health, National Institute of Allergy and Infectious Diseases, Laboratory of Pathobiology, Rocky Mountain Laboratories, Hamilton, MT 59840, USA*

Introduction

Relapsing fever is a distinctive disease that has been known to physicians since the time of Hippocrates. Recurrent febrile periods, which are measured in days and are spaced by week-long intervals of well-being, characterize the disease (*Figure 1*).

In its epidemic form relapsing fever has affected millions of people, and, like plague, often accompanies wars, famines and social upheavals (reviewed by Felsenfeld, 1971). The agent of epidemic relapsing fever, *Borrelia recurrentis*, is transmitted by lice, and man appears to be the only mammal host in nature. *B. recurrentis* has not been successfully established in small laboratory animals or cultivated serially *in vitro* and, as a consequence, these organisms have not been extensively studied. Instead, most investigators of borreliae rely on one of the many species of *Borrelia* that cause endemic relapsing fever. These types of borreliae are carried by ticks and primarily occur in reservoirs of rodents or other wild mammals. Man is not a necessary part of the cycle of transmission, but when endemic disease does occur in humans, it resembles the epidemic form in clinical features.

The tick-borne species can be passed transovarially through several generations of ticks and yet still retain virulence for mammalian hosts. Little is known of borreliae as they exist in their arthropod habitats. Attention has mostly been on infections of mammalian bloodstreams and tissues; presumably these places pose very different environments from those found in ticks and lice. For one thing, arthropods do not have the immunoglobulin-based defence system that vertebrates use.

Observers have long recognized that borreliae sampled from the blood at the time of the first relapse differed both from the infecting population and the succeeding relapse populations. Meleney (1928) succinctly summarized the state of knowledge about the relapsing phenomena a half century ago: 'At the time of the crisis which terminates the attack of fever, there is rapid agglutination and destruction of the spirochetes with the subsequent formation of immune bodies in the blood. These substances are specific for the strain of spirochetes which was present during the preceding attack, but have

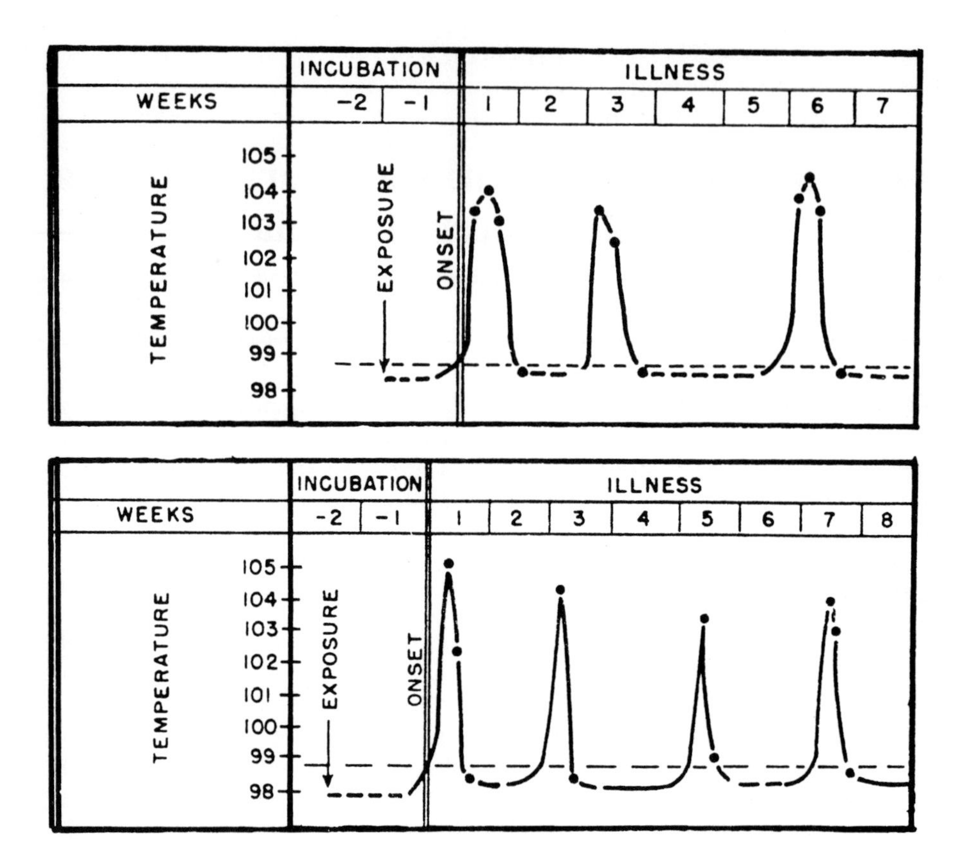

Fig. 1. Fever curves in two human cases of tick-borne relapsing fever. Body temperatures are given in degrees Fahrenheit. (After Thompson *et al.*, 1969.)

no influence on the spirochetes of the succeeding relapse. The spirochetes of the relapse give rise, in turn, to immune substances which are specific for them but not for the spirochetes of the first attack'. Antigenic variation has thwarted efforts — perhaps predictably — to produce an effective vaccine against relapsing fever borreliae (Felsenfeld, 1976).

In the decades following Meleney's paper, relapsing fever attracted relatively little attention. Its scientific obscurity could be attributed in part to difficulties in obtaining large enough numbers of the borreliae for adequate study. The discovery that borreliae replicated *in vitro* when media were supplemented with N-acetylglucosamine — coincidentally or not, the building block of arthropod chitin — permitted physiological, biochemical and genetic investigations of these organisms (Kelly, 1971).

Biology of relapsing fever borreliae

Borreliae are membranes of the eubacterial 'phylum' of spirochetes (Fox *et al.*, 1980). Borreliae share with other spirochetes such distinguishing structural characteristics as a cork-screw shape and internal flagellae that traverse the periplasmic spaces (Holt, 1978). Spirochetes, like the Gram-negative group of bacteria, have both cytoplasmic and outer membranes. However, the spirochetal outer membrane is more easily released from its underlying attachments and appears to be more fluid in constitution than its Gram-negative bacterial counterpart (Holt, 1978; Klaviter and Johnson, 1979; Barbour *et al.*, 1983b). The consistency of the spirochete's outer membrane probably determines the type of proteins that can be inserted in or through it.

Schuhardt and Wilkerson (1951) first demonstrated that a single borrelia sufficed to infect rodents, and Stoenner *et al.* (1982) utilized this valuable attribute to clone, by limiting dilution, a strain of *Borrelia hermsii* in mice. Stoenner also found that an *in vitro* culture begun with one organism was possible. Kelly's original medium was modified to achieve this and, using cloned populations, Stoenner developed serotype-specific antisera. Previous workers who looked at the relapsing fever phenomenon in laboratory animals were undoubtedly working with heterogeneous populations of organisms.

Through repeated cycles of first cloning the spirochetes, then producing antiserum to the cloned population, and finally examining the relapse populations for those that did not react with any of the previously produced antisera, Stoenner isolated 25 distinct serotypes from the progeny of the original infecting borrelia. By indirect immunofluorescence assays there was seldom evidence of cross-reactivity between the serotypes, and consequently the bank of 25 antisera could be used confidently to follow the courses of infection in mice and rats. Development of monoclonal antibodies to some of the serotypes provided complementary reagents for distinguishing between serotypes. Antigenic variation in mice and in culture media may be summarized as follows.

(i) Although any of the 25 serotypes may occur in the first relapse, there is a higher prevalence of certain serotypes, such as those designated numbers 7 and 2, during the early phases of infection (Stoenner *et al.*, 1982; Barbour and Stoenner, 1984; *Figure* 2). Those serotypes, such as 5 and 8, that are less common during the first or second relapses seem to come to the fore in the later relapses.

(ii) The appearance of new serotypes during infection or *in vitro* cultivation is spontaneous, i.e. the presence of serotype-specific antibody is not necessary for the switch to occur. Needless to say, neutralizing antibody is the paramount selective pressure applied by the mammalian host to borrelial populations undergoing variation.

(iii) Those organs, such as brain and eye, that are immunologically sequestered can harbour the originally-infecting serotype at a time when this inoculated serotype has long been absent from the blood (H.G.Stoenner, personal communication).

(iv) The rate of appearance of new variants or serotypes *in vivo* and *in vitro* has been estimated to be 10^{-4} to 10^{-3} per cell generation (Stoenner *et al.*, 1982).

(v) Change in serotypes is reversible. In other words, a relapse population, which has been recovered from a mouse at a time when it is immune to the infecting serotype, can be cloned and injected into a second, non-immune mouse and, during the course of infection in the second mouse, the relapse serotype serves as the source of the original infecting serotype. The information encoding the serotype-specific antigens has not, therefore, been lost during the variation process.

(vi) Polyclonal antisera or monoclonal antibodies that are serotype-specific can be used to passively 'select out' new serotypes in mice and in broth cultures (Stoenner *et al.*, 1982; Barbour and Stoenner, 1984).

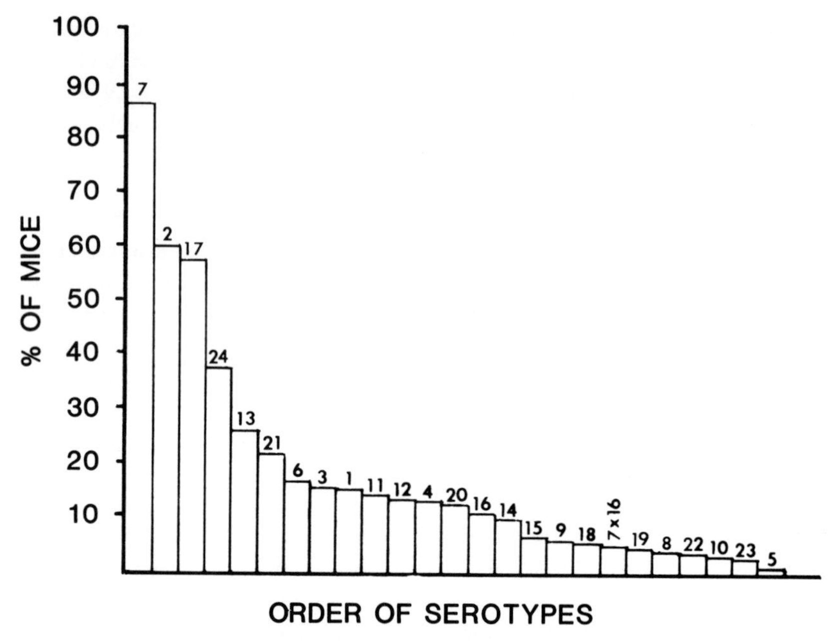

Fig. 2. Frequency of different serotypes of *B. hermsii* strain HS1 in mice undergoing first relapses of spirochetaemia. Data are taken from Table I of Barbour and Stoenner (1984). Serotype 7 was the most frequent serotype found during first relapses: 87% of mice had detectable concentrations of this serotype. '7 × 16' was a new serotype that reacted with both anti-serotype 7 and anti-serotype 16 sera (Stoenner *et al.*, 1982).

(vii) Some serotypes have a growth advantage over other serotypes in broth cultures and, because of this, predominate during *in vitro* cultivation. One such serotype, C, can produce a spirochetaemia in mice, but relapses due to other serotypes have not been detected in serotype C-infected mice (A.Barbour, unpublished observations). In the course of the switch to serotype C, genetic information may indeed have been lost from the cells.

Variable antigens of *Borrelia hermsii*

Serotype identity is conferred by abundant proteins that are on the surface of the spirochete (Barbour *et al.*, 1982, 1983a; Barstad *et al.*, 1985). *Figure 3* shows these proteins, which have been designated variable major proteins (VMPs) in five serotypes of the HS1 strain of *B. hermsii*. The other major protein in the polyacrylamide gel shown is, unlike the VMPs, invariable in its apparent molecular weight of about 40 000 among the different serotypes. This second protein, the pII protein (Barbour *et al.*, 1982), appears to be a structural protein of the periplasmic flagella (A.G.Barbour, in preparation).

Not only do the VMPs differ in molecular weights, they also contain the serotype-specific epitopes (Barbour *et al.*, 1982, 1983a; Barstad *et al.*, 1985). Monoclonal antibodies that only recognize one or another of the 25 different serotypes in immunofluorescence assays bind to the homologous VMP but not to heterologous VMPs in Western blots.

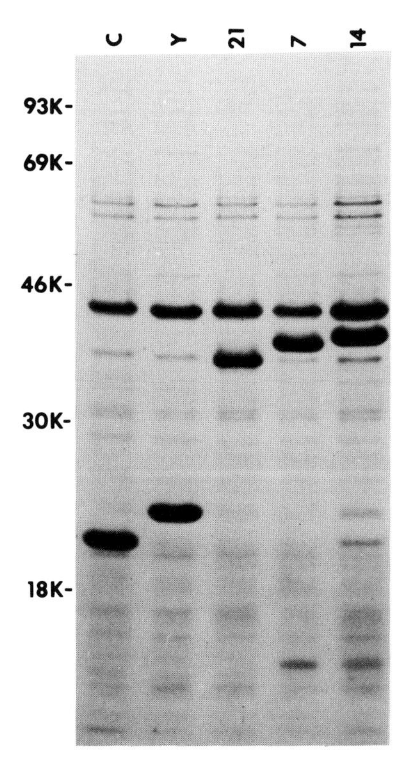

Fig. 3. Coomassie blue-stained polyacrylamide gels of *B. hermsii* HS1 serotypes C, Y, 21, 7 and 14 (Barbour and Stoenner, 1984). Each serotype has two major proteins: an invariable protein with an apparent molecular weight of 40 000 (40 K) and one of the proteins that differ in apparent molecular weight between serotypes. The latter proteins are the VMPs. The locations of the relative migrations of molecular weight standards are shown on the left.

The surface exposure of the VMPs was demonstrated by the following experiments.

(i) *In situ* VMPs were cleaved from the cell by proteases (Barbour, 1985).

(ii) When live cells were used, VMPs were radioiodinated under surface-specific labelling conditions (Barbour *et al.*, 1982).

(iii) VMP-reactive monoclonal antibodies bound to and agglutinated homologous borreliae (Barbour, 1985).

Although the VMPs differed in their sizes, *a priori* it was possible that VMPs contained a large constant region as is the case for gonococcal pili (Schoolnik *et al.*, 1982) and for immunoglobulins (Early *et al.*, 1979). Peptide mapping studies suggested that there were no extensive constant regions in the VMPs, however; three VMPs had few if any chymotrypsin or V8 protease peptides in common (Barbour *et al.*, 1983a).

Two VMPs were studied in greater detail (Barstad *et al.*, 1985). One VMP was isolated from serotype 7, the variant that is most often among the first relapse populations (*Figure 2*). The other VMP was that of serotype 21, a less frequent serotype during early infection but one which had been derived from a serotype 7 cell line (Fox *et al.*, 1980). *Figure 4* summarizes data relating to the primary structures of these two VMPs.

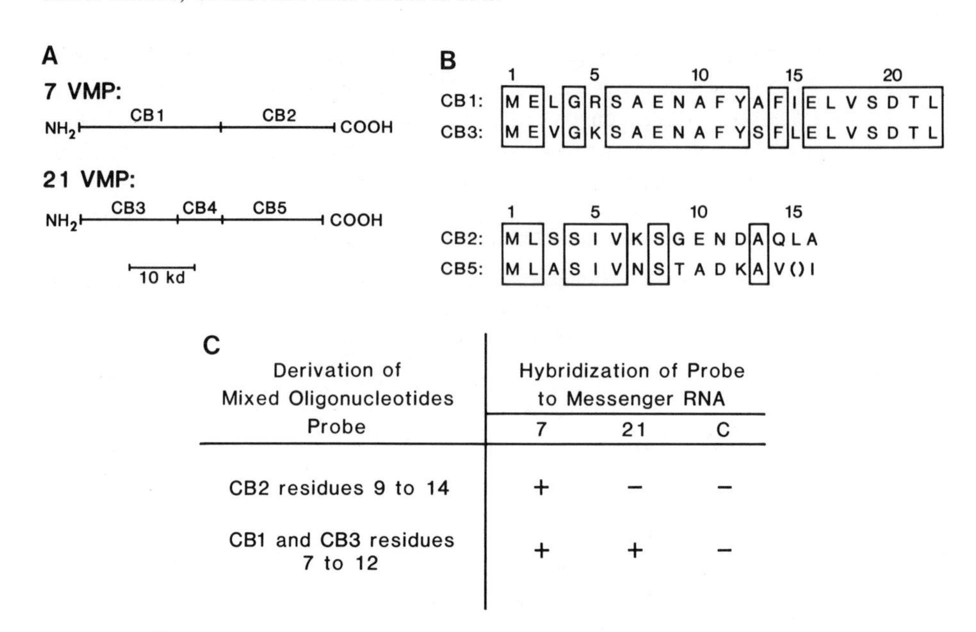

C

Derivation of Mixed Oligonucleotides Probe	Hybridization of Probe to Messenger RNA		
	7	21	C
CB2 residues 9 to 14	+	−	−
CB1 and CB3 residues 7 to 12	+	+	−

D

Monoclonal Antibody	Reactivity of Antibody							
	Whole Cells and VMPs			CNBr Fragments				
	7	21	C	CB1	CB2	CB3	CB4	CB5
A1	+	−	−	−	+			
A2	+	−	−	+	−			
A3	+	+	−	+	−	−	−	+
A4	−	+	−			−	−	+
A5	−	+	−			−	+	−
A6	−	+	−			+	−	−

Fig. 4. **(A)** Cyanogen bromide-generated fragments of VMP proteins of serotypes 7 and 21 (Barstad *et al.*, 1985; S.Bergstrom and A.Barbour, in preparation). Bar, molecular weight of 10 000 (10 kd). **(B)** Amino acid sequence analysis of the N termini of four CNBr fragments of 7 VMP and 21 VMP (Barstad *et al.*, 1985). Sequences with partial homologies were paired and aligned. Single letter abbreviations are as follows: A, alanine; D, aspartic acid; E, glutamic acid; F, phenylalanine; G, glycine; I, isoleucine; K, lycine; L, leucine; M, methionine; N, asparagine; P, proline, Q, glutamine; R, arginine; S, serine; T, threonine; V, valine; W, tryptophan; and Y, tyrosine. Boxes indicate regions of homology between pair members. **(C)** Hybridization of oligonucleotide probes to mRNA from serotypes 7, 21 and C (Meier *et al.*, 1985). Mixed oligonucleotides were derived from amino acid sequences that were unique to 7 VMP (CNBr2 residues 9−14) or common to both 7 VMP and 21 VMP (CNBr1 and CNBr3 residues 7−12). Plus (+) sign indicates hybridization of probe to mRNA in Northern blots. **(D)** Reactivities of monoclonal antibodies A1−A6 with whole cells or intact VMPs of serotypes 7, 21 and C and with CNBr fragments of 7 VMP and 21 VMP (derived from Table I of Meier *et al.*, 1985). Whole cell reactivity was examined with indirect immunofluorescence assays; specificities of antibodies for intact VMPs and CNBr fragments of VMPs were determined by Western blot analysis. Antibody A3 bound to cells, proteins and peptides of serotypes 7 and 21 but not of serotype C.

The 7 VMP has two CNBr fragments; the 21 VMP has three (*Figure 4A*). Partial N-terminal amino acid sequencing of the five CNBr fragments (CNBr1 – CNBr5) revealed sequence homologies between two pairs of fragments: CNBr1/CNBr3 and CNBr2/CNBr5 (*Figure 4B*). There is a greater homology (~80%) between the two N-terminal regions of each protein than between those regions, i.e. CNBr2 and CNBr5, that are closer to the C termini. Only 40% homology exists between the aligned sequences of CNBr2 and CNBr5. Moreover, half of the amino acid replacements in these two sequences would have required a change of at least two bases in the corresponding codons. Thus, there appears to be a considerable evolutionary distance between some parts of VMPs.

When two mixed oligonucleotide probes that were based upon amino acid sequences of 7 and 21 VMPs were reacted with total mRNA obtained from each of these serotypes as well as mRNA from serotype C, the probe that was generated from a peptide sequence of CNBr2 bound only to mRNA from serotype 7 (*Figure 4C*) (Meier *et al.*, 1985). On the other hand, the probe derived from a peptide that was common to both VMPs hybridized to mRNA from serotypes 7 and 21 but not to mRNA from serotype C. This showed that 7 and 21 VMPs had in common regions of sequence that were not shared with C VMPs and that serotypes could be distinguished at the transcriptional level. The latter conclusion was supported by the observation that a borrelial DNA restriction fragment specific for the 21 VMP gene hybridized only to an mRNA species found in serotype 21 cells (Plasterk *et al.*, 1985).

Isolation of each of the five CNBr fragments allowed not only sequence anlaysis but also epitope mapping with a battery of monoclonal antibodies to serotypes 7 and 21 (*Figure 4D*). Monoclonal antibodies were specific for either serotype 7 or 21 with the exception of antibody A3 which reacted with cells and VMPs of both 7 and 21. (Antibody A3 does not react with any of the other serotypes that have been examined.) The cross-reactivity of antibody A3 provides additional evidence that 7 and 21 VMPs have structures in common. The other monoclonal antibodies that are listed are serotype-specific in their reactivities, and demonstrate that serotype-specific epitopes are dispersed throughout the length of the VMP molecules. The lack of confinement of serotype-specific epitopes to one region is further indication that there is not an extensive constant region. Instead, the antibody reactivities, the peptide maps and the sequence data obtained to date suggest that there are comparatively small regions of close amino acid sequence homology alternating with regions of heterology. It seems likely, therefore, that the VMPs belong to one or more polygene families (Hood *et al.*, 1975), as do the variable surface proteins of the trypanosomes (Lalor *et al.*, 1984). We can speak, therefore, of *B. hermsii* as being polymorphic with respect to VMPs.

Activation of transcription of VMP genes is associated with DNA rearrangements

The biology of the relapsing phenomena, the polymorphic nature of VMPs and the differential transcription of VMP genes indicated that antigenic variation in borreliae is the result of hereditary changes. To determine if the hereditary difference between a serotype 7 cell and cells of other serotypes involved rearrangements of the genetic material, a DNA sequence that was specific for the 7 VMP gene was used to probe Southern blots of restriction enzyme-digested DNA (Meier *et al.*, 1985). The 7 VMP-

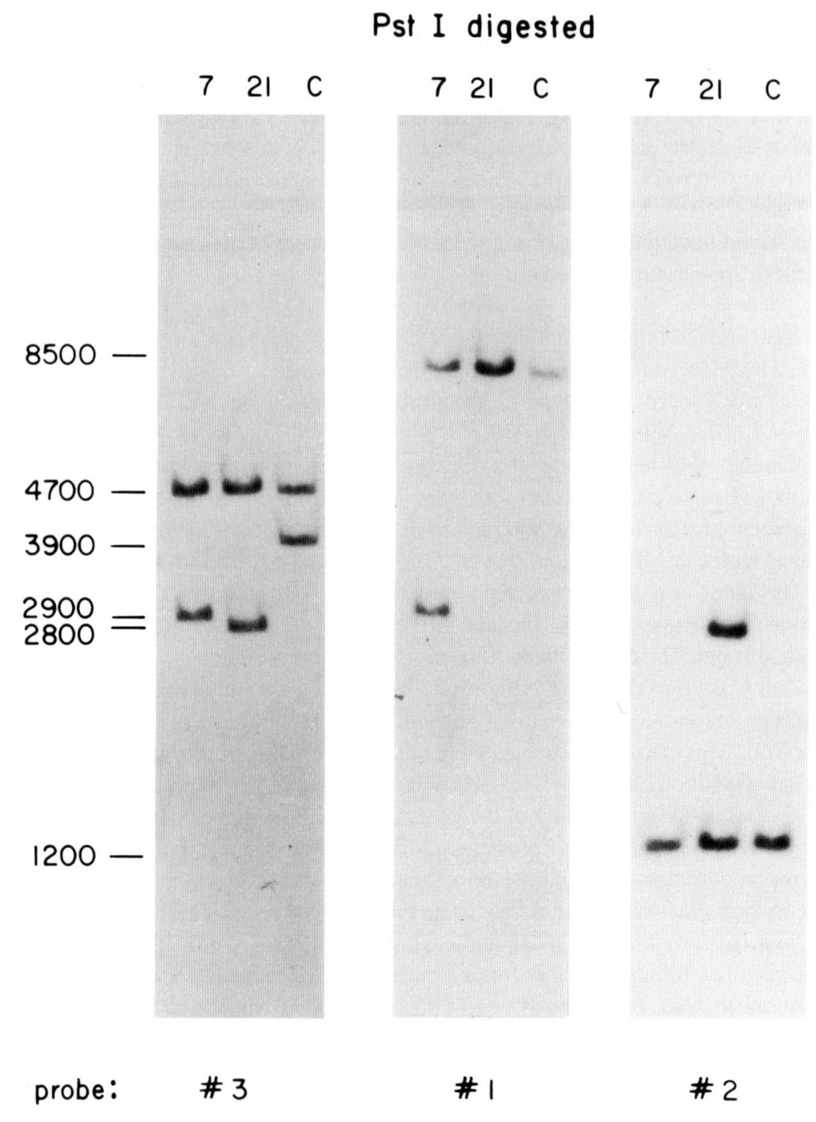

Fig. 5. Southern blot analysis of *Pst*I-digested DNA from serotypes 7, 21 and C of *B. hermsii*. The probes were No. 3 (left panel), No. 1 (middle panel) and No. 2 (right panel). The No. 1 probe hybridized to an additional band of 2.9 kbp in the 7 DNA, the No. 2 probe hybridized to an additional 2.8-kbp fragment in the 21 DNA. Note that the No. 3 probe hybridized to a 2.9-kbp fragment in the 7 DNA, a 2.8-kbp band in the 21 DNA and a 3.9-kbp fragment in the C DNA. These are all three fragments that are specific for each serotype. In addition probe No. 3 hybridized to a fragment of 4.7 kbp present in all three serotypes. This *Pst*I fragment has been mapped; the plasmids p7.7 and p21.4 (i.e. the plasmids containing the expression site) contained a 4.7-kbp *Pst*I fragment that hybridized with the probe No. 3 (not shown). This *Pst*I fragment was mapped adjacent to the *Pst*I fragment that contains the recombination site (see *Figure 7A*).

specific probe hybridized to a restriction fragment common to all serotypes and to an additional fragment only in serotype 7 DNA. Subsequently, probes for the 21 VMP gene as well as the 7 VMP gene were used in Southern blot analysis of DNA from different serotypes (Plasterk *et al.*, 1985). *Figure 5* shows the pattern obtained when DNA of serotypes 7, 21 and C is digested with the restriction enzyme *Pst*I and hybridized with a probe specific for the 7 VMP gene or a probe specific for the 21 VMP gene. The serotype 7-specific probe hybridizes to a band common to all three serotypes and to an additional band in the serotype 7 DNA. In a similar fashion, a DNA sequence from the 21 VMP gene hybridizes to one band in 7 and C DNA and to two bands in 21 DNA. From these experiments we concluded that the activation of 7 and 21 VMP transcription is associated with the appearance of 7 and 21 VMP DNA sequences in a new restriction fragment and, therefore, that antigenic variation is associated with DNA rearrangements. The findings also suggested that the extra copy of the VMP gene sequence is the copy that is transcribed.

Silent and expressed VMP genes

To test the proposition that an extra copy of the 7 or 21 VMP gene is responsible for the expression of that gene, both copies of these genes, i.e. the copy common to all serotypes and the copy specific for either the 7 or 21 serotype, were cloned (Plasterk *et al.*, 1985). Briefly, this was done as follows. DNA from serotype 7 or 21 cells was partially digested with the enzyme *Sau*3A and cloned into a bacteriophage lambda cloning vector. Recombinant phage clones containing the 7 and 21 VMP genes were selected from the gene libraries using probes that were specific for either the 7 or 21 VMP genes. The inserts of the identified phages were subcloned into the plasmid pBR322 and the resulting recombinants were transformed into *Escherichia coli*. Western blot analysis with monoclonal antibodies to 7 and 21 VMPs determined which of the clones were synthesizing immunoreactive products and, thus, established the state of VMP gene expression in each of the clones. *Figure 6* shows the result of one such analysis: some of the clones that contain the 7 VMP gene express the protein at a high level, and some do not at all. Similarly, some 21 VMP gene-containing clones express 21 VMP, and some do not.

The structure of the silent and expressing clones was examined by restriction enzyme mapping (Plasterk *et al.*, 1985). The results are shown in *Figure 7*. Although only representative examples are shown, the restriction maps of the sequences around the silent 7 VMP genes were identical in all clones examined. This was also true for the set of clones containing the expressed 7 VMP gene. However, when the silent and expressing clones of the 7 VMP gene were compared with one another, it was apparent that the DNA sequences upstream, i.e. to the left in the figure, of the VMP gene were completely different in the two types. It can be concluded that transcription of the 7 VMP gene is associated with the presence of a particular DNA sequence flanking one of the copies. Whereas the clones of the expressed 7 VMP gene contained the restriction fragment that was specific for serotype 7 (see *Figure 7*), the silent clones contained the restriction fragment that was common to all serotypes, i.e. when the 7 VMP gene present in all serotypes was cloned into *E. coli*, 7 VMP expression did not occur. In contrast, a clone containing the restriction fragment specific for serotype 7 did express the protein. This indicates that the extra copy of the 7 VMP gene in serotype 7 is the one transcribed, and that the other copy is the silent one. Such was also the case for the silent and expressing 21 VMP clones: both types of clones had characteristic restriction maps, and the fragment unique to serotype 21 was contained in the clone that ex-

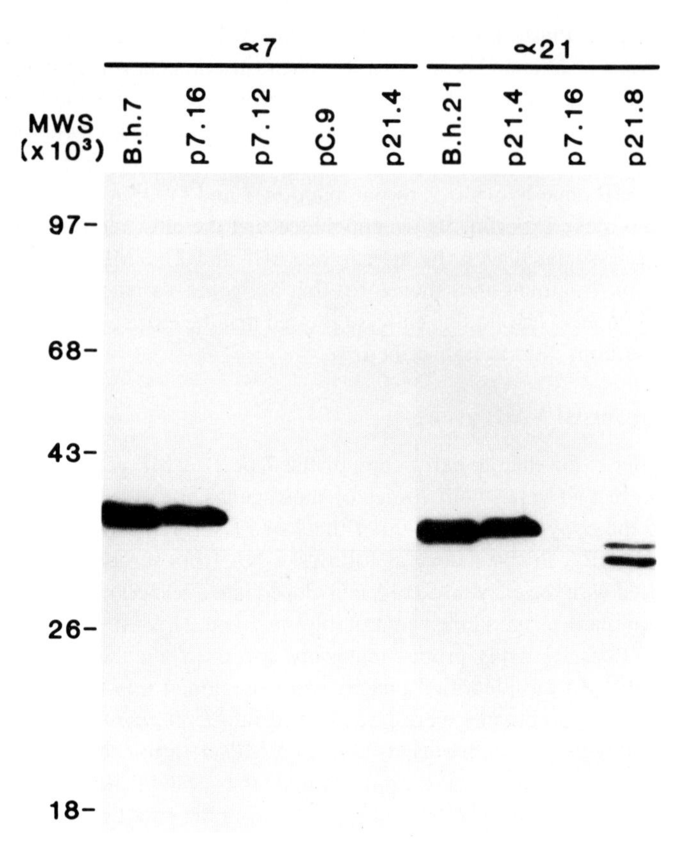

Fig. 6. Western blot analysis of lysates of *B. hermsii* or of *E. coli* containing recombinant plasmids with *Borrelia* DNA. Monoclonal antibody H12436 (Meier *et al.*, 1985) is anti-7 VMP (α7; left panel); monoclonal antibody H10022 (Meier *et al.*, 1985) is anti-21 VMP (α21, right panel). Lanes contained lysates of *B. hermsii* serotypes 7 (B.h.7) or 21 (B.h.21) or lysates of *E. coli* JM101 containing p7.16, p7.12, pC.9, p21.4 or p21.8. On the left are shown the relative migrations of molecular weight standards (MWS).

presses 21 VMP. The recombinant clones that did not produce 21 VMP contained the fragment common to all serotypes examined. In summary, while all serotypes contain silent copies of the 7 and 21 VMP genes, an extra copy of a given VMP gene is found only in the cell type in which that VMP gene is expressed.

The expressed VMP genes are fused to an expression sequence

The expressed copies of the 7 and 21 VMP genes are flanked at their 5′ ends by sequences that are different from those flanking the silent VMP genes. From a comparison of the restriction maps of the DNA sequences upstream of the expressed 7 and 21 VMP genes, we concluded that these upstream flanking regions were likely to be identical (Plasterk *et al.*, 1985) and the identity between upstream sequences was confirmed by Southern blot analysis. Inasmuch as only the VMP genes that are fused to the sequence in question are transcribed, this DNA sequence can be referred to as the expression sequence. A DNA fragment internal to the expression sequence was used, in turn, as

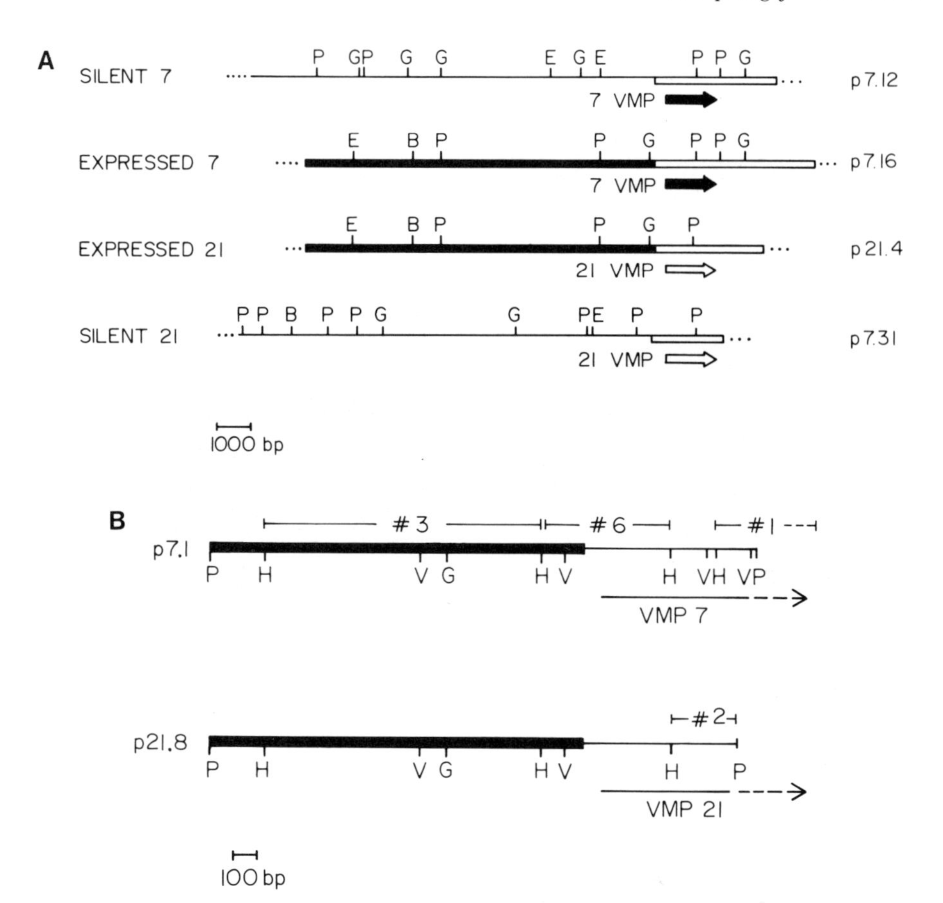

Fig. 7. **(A)** The silent and expressed 7 and 21 VMP genes and surrounding sequences. Restriction maps are shown of the clones p7.12 (silent 7 VMP from the 7 gene bank), p7.7 and p7.16 (overlapping clones containing expressed 7 VMP), p21.4 (expressed 21 VMP) and p7.31 (silent 21 VMP from the 7 gene bank). The sites indicated are: P + *Pst*I, G = *Bgl*II, B = *Bam*HI, E = *Eco*RI. The 1.9-kbp *Pst*I fragment downstream of the expressed 21 VMP gene (in p21.4) was used as a probe in Southern analysis of p7.16, and no homology was detected (not shown). This means that in the DNA segments shown here the sequences downstream of the expressed 7 and 21 VMP genes are not identical (Plasterk *et al.*, 1985). **(B)** The expressed VMP genes and the expression sequence. Shown are the restriction maps of the 2.9- and 2.8-kbp *Pst*I fragments that are subclones of p7.16 and p21.4, respectively. These fragments contain the 5′ segment of the 7 and 21 VMP genes and part of the sequence upstream of these genes. These fragments have been cloned into the *Pst*I site of pBR322 (resulting in the clones p7.1 and p21.8). The letters indicate restriction sites: P = *Pst*I, H = *Hind*III, G = *Bgl*II, V = *Pvu*II. The heavy line indicates the DNA that is homologous in both restriction fragments. The thin lines indicate the DNA that is specific for either the 7 or the 21 serotype. The 7 VMP and 21 VMP open reading frames are indicated by a horizontal arrow. The restriction fragments that were used as probes in subsequent hybridization experiments were designated No. 1 (in the 7 VMP structural gene), No. 3 (in the expression sequence) and No. 2 (in the 21 VMP structural gene).

a probe in Southern blot analysis (*Figure 5*, left panel). The blot showed that the expression sequence was fused to the expressed 7 VMP gene in serotype 7 and to the expressed 21 VMP gene in serotype 21. In serotype C, it was fused to yet another sequence, possibly the expressed C gene, but this has not been proven. Alternatively, the C VMP gene may be activated by fusion to another, as yet unidentified, expression

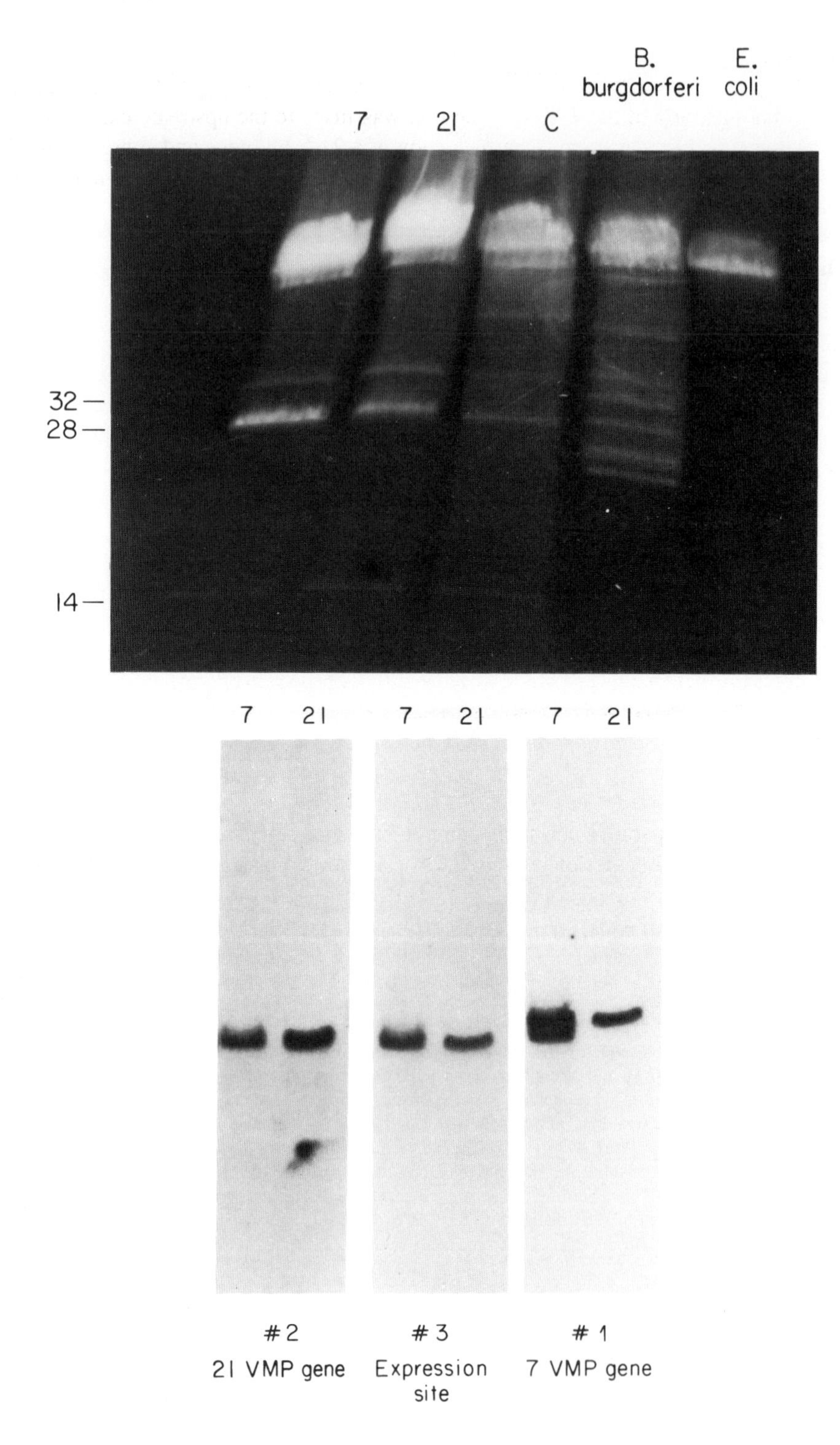
B. burgdorferi
E. coli
7
21
C
32—
28—
14—
7
21
7
21
7
21
2
21 VMP gene
3
Expression site
1
7 VMP gene

sequence.

In the particular switch from serotype 7 to serotype 21 that was observed, it appears likely that the copy of the 7 VMP gene that was fused to the upstream expression sequence was replaced at some point by a copy of a 21 VMP gene and consequently the 21 VMP gene became activated. The silent 7 and 21 VMP genes seem not to have been affected by the switch.

The VMP genes are on linear plasmids

The DNA of *B. hermsii* can be resolved into several discrete bands in standard, low-percentage agarose gels and on the pulse field gels of Schwartz and Cantor (1984). *Figure 8A* shows the pattern obtained when total DNA from *B. hermsii* serotypes 7, 21 and C was subjected to electrophoresis in a pulse field gel. The sizes of the ethidium bromide-stained bands from serotype 7 and 21 DNA appeared identical with a topmost band of chromosomal DNA, a plasmid of about 100 kb immediately beneath the chromosomal DNA band and plasmids of 32 kb, 28 kb and 14 kb. DNA from serotype C contained the same bands and an additional band of approximately 50 kb.

Extrachromosomal DNA resolvable in pulse field gels is not unique to *B. hermsii*. *B. burgdorferi*, the spirochetal agent of Lyme disease, contains numerous plasmids (*Figure 8A*) (Barbour *et al.*, 1984). In contrast, an *Escherichia coli* K12 strain (*Figure 8A*) and an isolate of *Spirocheta aurantia*, a non-pathogenic and free-living spirochete, did not have detectable plasmids when their DNA was examined in a similar way (R.H.A. Plasterk and A.G.Barbour, unpublished observations).

The VMP genes and the expression sequence were mapped on these borrelial plasmids by Southern blot hybridization with probes specific for the 7 and 21 VMP genes and for the expression sequence. *Figure 8B* shows the Southern blot analysis of serotype 7 and 21 DNA separated into their components on a standard 0.5% agarose gel. In both serotypes the expression sequence was present on a plasmid of 28-kb (*Figure 8B*, centre panel) but, in the 21 DNA, a 7 VMP gene — presumably the silent copy — was on a plasmid of 32 kb (*Figure 8B*, right panel). Also, the 7 DNA has the silent copy of the 7 VMP gene on a 32-kb plasmid and another copy of the 7 VMP gene on the 28-kb plasmid, the latter plasmid also containing the expression sequence. The left panel of *Figure 8B* shows the Southern blot pattern obtained using a probe specific for the 21 VMP gene: the hybridizing bands are associated only with the 28-kb plasmids of serotypes 7 and 21.

Fig. 8. (A) Pulsed field gel electrophoresis of *Borrelia* DNA. DNA from *B. hermsii*, *B. burgdorferi* and *E. coli* was isolated as described (Meier *et al.*, 1985) and separated as follows: a 1.1% agarose gel was placed in a tray with two pairs of electrodes [a slight modification of the original method described by Schwartz and Cantor (1984)]; electrophoresis was for 16 h at 175 V, with the field changing every second. The gel was stained with ethidium bromide and photographed. The top band is the chromosomal DNA (in lane 5 *E. coli* chromosomal DNA is shown). In addition there are several extrachromosomal DNA elements visible. The pulse time of 1 sec was chosen to optimize separation of the 28-kbp band from the 32-kbp band. When longer times are chosen the larger elements are better separated, and it is then clear that the band immediately below the chromosomal band is indeed a separate band. **(B)** Southern blot analysis of the extrachromosomal *B. hermsii* DNA. Total DNA from serotypes 7 and 21 was separated on a 0.5% agarose gel (25 V, 60 h at 4°C). The resulting pattern is identical to the one shown in *Figure 4A*. The same blot was used with the three probes, No. 1, No. 2 and No. 3.

There is evidence, however, that the silent 21 VMP gene is on a plasmid of approximately 28 kb that is distinct from the 28-kb plasmid containing the expressed copy of the gene. A recombinant clone with an insert of 15 kb that included the silent 21 gene (p7.31 of *Figure 7A*) did not have discernible homology with two overlapping recombinant clones each containing the expressed 7 VMP gene and, when taken together, extensive regions upstream and downstream from the gene. Thus, the silent and expressed copies of the 21 VMP gene are probably not parts of the same 28-kb plasmid.

One may conclude from the above experiments that *B. hermsii* contains plasmids carrying the silent VMP genes and a plasmid carrying both the expression sequence and the extra copy of a VMP gene. Additional experiments demonstrated that the plasmids containing the VMP genes are linear. The evidence for this statement is as follows.

(i) Circular DNA cannot be well resolved in the pulse field gel (Schwartz and Cantor, 1984) but *B. hermsii* plasmids, in contrast, are well separated with this apparatus.

(ii) After gentle lysis of the borreliae with non-denaturing detergents and extraction of the DNA without phenol or proteases, linear DNA of 20–30 kb can be observed by electron microscopy (A.G.Barbour and C.Garon, unpublished observations).

(iii) The DNA of the 28-kb plasmid has specific ends and is not circularly permuted. This was shown when total DNA from serotype 7 was treated in a time series with the double strand exonuclease, *Bal*31, digested with *Pst*I and the fragments separated on a gel. To identify fragments of the 28-kb plasmid, the entire 28-kb plasmid was isolated, nick-translated and used as a probe of a Southern blot of the twice-digested DNA. *Figure 9* shows that the probe bound to a limited number of bands at time zero (lane 1) and subsequent samples from the series showed that upon treatment with *Bal*31 one band decreased in molecular weight but other bands were unaffected. When the first affected band was totally degraded, a second fragment (indicated by arrow 'II') began to decrease in size. Only one of the specific ends is susceptible to *Bal*31 attack, the other end being protected, possibly by a protein or some other covalent modification of the terminal DNA.

Model for the molecular basis of antigenic variation in *Borrelia hermsii*

On the basis of the analysis of cloned VMP genes, of Southern blot hybridizations of *B. hermsii* DNA from different serotypes and of studies of the extrachromosomal DNA, Plasterk *et al.* (1985) proposed a model (*Figure 10*) for the antigenic switches in the relapsing fever borrelia, *B. hermsii*. It must be emphasized that this model originates from experiments involving only three serotypes and may not necessarily hold true for the other 22 serotypes of *B. hermsii* HS1 or for other species of *Borrelia*.

It is proposed that each serotype contains silent copies of all VMP genes with a different gene for each of the serotype-determining VMPs; such VMP genes are members of one or more polygene families. The silent genes are on linear plasmids of relatively low copy number, but cells of each serotype also contain an extra copy of one of the VMP genes, this extra copy being fused to the expression sequence on a plasmid of higher copy number. The second VMP gene is the one that is transcribed and a switch from one serotype to another is the consequence of a site-specific recombination reaction between a site at the border of the expression sequence and a site upstream of one

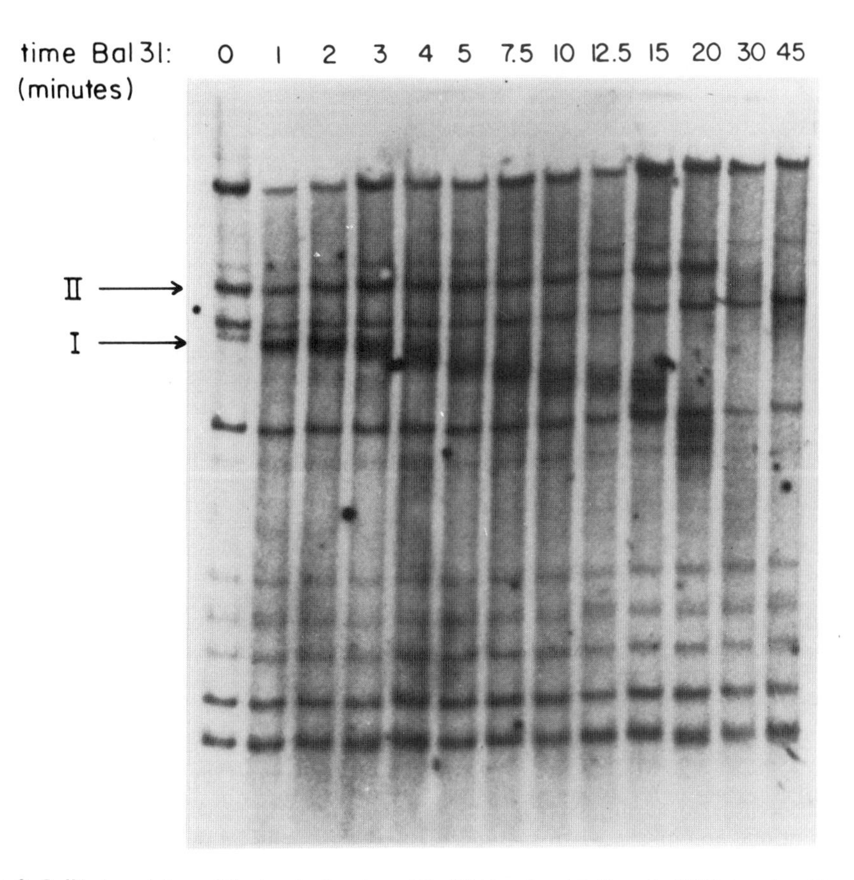

Fig. 9. *Bal*31 degradation of the terminal region of the 28-kbp plasmid. *Borrelia* DNA (serotype 7) was treated for the indicated time with 0.15 units of *Bal*31 (Plasterk *et al.*, 1985). Then EDTA was added to a final concentration of 20 mM and the DNA was recovered by alcohol precipitation, digested with *Pst*I, and separated on a 1% agarose gel. After blotting, it was hybridized with the '28-kbp probe'. This probe was obtained as follows: serotype 7 DNA was separated in a pulse field gel; after ethidium bromide staining, the 28-kbp plasmid band was cut out of the gel, and the DNA recovered and nick-translated. The arrow 'I' indicates the first band that disappeared; the arrow 'II' the second band.

of the silent genes. During cell division the newly formed expression plasmid is segregated from the other product of the recombination reaction and from the other expression plasmids. (The segregation could be passive or the result of a mechanism in the cell that distinguishes newly formed expression plasmids and existing expression plasmids.) After segregation all expression plasmids contain the new VMP gene.

Linear plasmids and pathogenic bacteria

To date, linear plasmids have been observed in some fungi (Francou, 1981) and in one other bacterium, a *Streptomyces* sp. (Hirochika and Sakaguchi, 1982). Linear intermediates are also present in bacteria during some stages of the life cycle of many types of bacteriophage. For most bacteriophages, the actual replication form is a covalently closed circle. An exception is ϕ29, a virus of *Bacillus subtilis*, which replicates as a

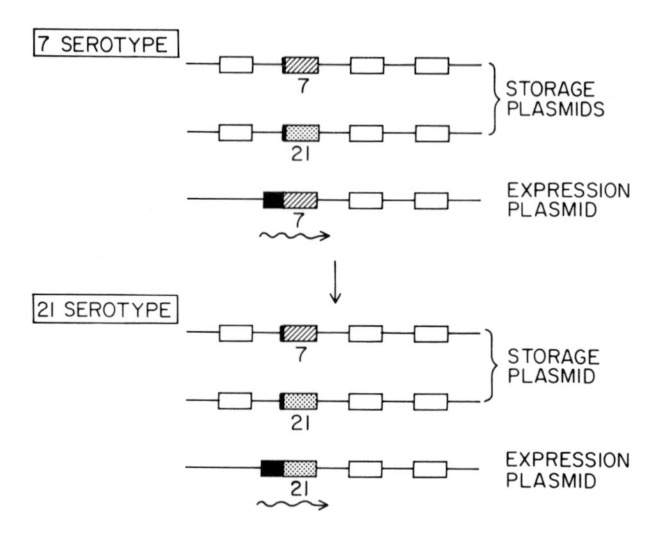

Fig. 10. Model for the molecular basis of antigenic variation in *B. hermsii*. Shown are the relevant plasmids of the 7 and 21 serotypes. The storage plasmids are identical in both strains. The 21 VMP gene in the serotype 7 cells is not transcribed. It transposes into the expression site, and the resident 7 VMP gene is removed. The VMP gene that is adjacent to the expression sequence is transcribed. As discussed in the text, several mechanisms for the recombination reaction are possible, but we currently favour the model in which one reciprocal site-specific recombination takes place; this event is followed by loss of one of the two reaction products.

linear molecule (Watabe *et al.*, 1982). DNA polymerization begins at proteins which are covalently bound to the termini of the DNA strands (Watabe *et al.*, 1984). The block at one end of the 28-kb plasmid of *B. hermsii* may be due to a protein which serves either to initiate DNA replication of a linear form or to bind together the two ends of the molecule into a circle.

There are several theoretical advantages to a pathogen in having its surface antigen genes arrayed on linear extrachromosomal DNA.

(i) The copy number of the plasmids can be high, and it is possible to change the copy number of a specific gene by a rearrangement. This occurs in *B. hermsii* where the number of copies of plasmids with the silent 7 and 21 VMP genes appears to be several times lower than that of the expression plasmid (see *Figure 8A*). The cell can efficiently store the silent genes in a low copy number plasmid and reserve the higher copy number plasmid for expression.

(ii) The plasmids can undergo rearrangement at a high frequency without disturbing the integrity of the chromosome.

(iii) There is the potential for horizontal transfer of genetic information in a cell population through transformation, conjugation or transduction. This would allow genes to be distributed comparatively rapidly throughout a species. It is not yet known whether any of these methods of gene transfer are possible in borreliae or other spirochetes.

(iv) Linear plasmids can recombine by a single reaction without the need for a resolution step. Between two circular plasmids a single cross-over event results in one large circular product and an additional recombination event is required to resolve this structure. Such constraints do not apply to linear DNA molecules.

The advantages of linear plasmids could also apply to other types of pathogenic bacteria and one such pathogen, *Neisseria gonorrhoeae*, has been studied with this in mind. The gonococcus, like borreliae, undergoes antigenic variation and transposition of pilus-specifying genes is a mechanism for this variation (Segal *et al.*, 1985). However, a probe for the pilin gene hybridized only to chromosomal DNA when total DNA from a strain of *N. gonorrhoeae* was examined (R.H.A.Plasterk and M.So, unpublished observations). Nonetheless, it may be useful to survey other pathogens to see if they contain virulence-associated genes on linear plasmids. Although standard methods of electrophoresis of superhelical plasmid preparations are unlikely to reveal linear plasmids, pulse field gel electrophoresis could serve to identify them.

Mechanisms of gene switching

Two mechanisms have been described for the activation of a gene through fusion of that gene to a specific DNA sequence. One mechanism is the placement of an insertion element upstream of the structural gene and the other is the transposition of the gene itself into an expression site. This latter mechanism occurs during the switch of mating type 'cassettes' in yeast (Hicks *et al.*, 1977) and the movement of antigen-encoding genes between different chromosomes of the African trypanosomes (Van der Ploeg *et al.*, 1984). In each of these cases a mobile DNA element inserts into another DNA sequence and in the process creates two new 'fusion sites',i.e. the two sites where DNA sequences that previously were not adjacent are now coupled. In the case of *B. hermsii* — specifically the switch from serotype 7 to serotype 21 — only one such fusion or recombination site seems to have been created. A fusion site can be detected with a probe that contains one or both of the sequences that are recombined when the patterns resulting from hybridization of this probe to digested DNA from before and after the switch are compared. Two fusions or recombination sites should yield two restriction fragment length polymorphisms. This was tested in *B. hermsii* with the 28-kb expression plasmid that had been nick-translated; when the probe was hybridized to restriction enzyme-cut DNA from serotypes 7 and 21, only one polymorphism was detected in Southern blots (R.H.A.Plasterk, unpublished observations). The polymorphic fragment contained the site of fusion of the VMP gene to the expression sequence and, as demonstrated above, restriction enzyme maps of the silent and expressing 7 and 21 VMP gene clones identify only one possible recombination site, i.e. upstream of the expressed copy of the VMP gene. In light of the linearity of the plasmids, a single recombination site for the 7 VMP to 21 VMP switch is conceivable (and, as discussed, teleologically advantageous). Although we cannot exclude the existence of other cross-over sites at a greater distance downstream than could be detected, a site-specific recombination at a single site seems most likely. This probability raises the question of which sequences upstream of the silent VMP genes signal a putative 'recombinase' to act to fuse that gene to the expression site. All these genes may be preceded by a short sequence that is more or less homologous to a sequence at the border of the expression sequence and that serves as a core in a site-specific recombination reaction. Different degrees of divergence from a consensus sequence for this site could account for the different prevalences of serotypes during early relapses; the silent 7 VMP gene may, for example, have upstream to it a core sequence that is more homologous to the consensus sequence than the silent VMP gene for an infrequent serotype like 5. Sequence analyses of silent VMP genes and their flanking regions should give answers to the questions that have been raised.

Concluding remarks

B. hermsii is well suited for study of the genetic mechanisms of antigenic variation in a pathogenic microorganism. There are several cogent reasons for this. First, the antigenic repertoire is elaborate: 25 variant serotypes have been derived from one serotype (Stoenner *et al.*, 1982; Barbour *et al.*, 1984). Second, the animal model of relapsing fever is simple and mimics the human disease; in addition, populations of borreliae are easily sampled. Third, infection in mice can be initiated by a single spirochete (Schuhardt and Wilkerson, 1951; Stoenner *et al.*, 1982). Fourth, the organism can be cultivated *in vitro* (Kelly, 1971). Fifth, the variable antigens are abundant surface proteins that differ extensively in their primary structures (Barbour *et al.*, 1982, 1983a; Barstad *et al.*, 1985). Sixth, antigenic variation has been observed *in vivo* and *in vitro* at a frequency of about 1 in 10 000 (Stoenner *et al.*, 1982). These properties permit isolation of cell populations of clonal origin, development of specific antibody and oligonucleotide probes, and analysis of genetic material before and after the antigenic change. As demonstrated by the experiments reviewed here, a complex genetic switch in this bacterium should soon be understood in functional, structural and mechanistic terms.

As stated earlier, these studies have examined only the mammalian portion of the cycle of transmission of *B. hermsii*; much less is known about these organisms in their arthropod habitats. Antigenic variation is clearly useful to borreliae in the mammalian bloodstream, but is it necessary in a tick which lacks anything as formidable as immunoglobulins? Perhaps the environmental changes that accompany the shift of the microorganism from arthropod to mammal might influence or conceivably even ‘trigger’ antigenic variation. As has been shown here, antigenic variation in *B. hermsii* is essentially synonymous with inheritable changes in the DNA.

François Jacob (1982) has written: ‘For modern biology there is no molecular mechanism enabling instructions from the environment to be incorporated into DNA directly, that is, without the roundabout route of natural selection. Not that such a mechanism is theoretically impossible. Simply it does not exist.’ It could be that an organism like *B. hermsii* that has two very different environments in which to live responds to specific signals from the environment by starting or stopping a series of DNA rearrangements.

Acknowledgements

We thank Paul Barstad, Sven Bergstrom, John Coligan, Ralph Judd, Leonard Mayer, Joe Meier and Herb Stoenner for their valued contributions and Betty Kester for expert preparation of the manuscript. A portion of the work reviewed here was supported in part by grant DMB 85-16021 from the National Science Foundation to M.I.S. and by a grant from the Netherlands Organization for the Advancement of Pure Research (Z.W.O.) to R.H.A.P.

References

Barbour,A.G. (1985) Clonal polymorphism of surface antigens in a relapsing fever *Borrelia* sp. In *Bayer Symposium VIII: The Pathogenesis of Bacterial Infections.* Jackson,G.G. and Thomas,H. (eds), Springer-Verlag, Heidelberg, pp. 235–245.

Barbour,A.G. and Stoenner,H.G. (1984) Antigenic variation of *Borrelia hermsii*. In *Genome Rearrangement. UCLA Symposia on Molecular and Cellular Biology, New Series, Vol. 20.* Herskowitz,I. and Simon,M. (eds), Alan R.Liss, Inc., New York, pp. 123–135.

Barbour,A.G., Tessier,S.L. and Stoenner,H.G. (1982) Variable major proteins of *Borrelia hermsii. J. Exp. Med.*, **156**, 1312–1324.

Barbour,A., Barrera,O. and Judd,R. (1983a) Structural analysis of the variable major proteins of *Borrelia hermsii. J. Exp. Med.*, **158**, 2127–2140.

Barbour,A.G., Tessier,S.L. and Todd,W.J. (1983b) Lyme disease spirochetes and ixodid tick spirochetes share a common surface antigenic determinant defined by a monoclonal antibody. *Infect. Immun.*, **41**, 795–804.

Barbour,A.G., Tessier,S.L. and Hayes,S.F. (1984) Variation in a major surface protein of Lyme disease spirochetes. *Infect. Immun.*, **45**, 94–100.

Barstad,P.A., Coligan,J.E., Raum,M.G. and Barbour,A.G. (1985) Variable major proteins of *Borrelia hermsii*: epitope mapping and partial sequence analysis of CNBr peptides. *J. Exp. Med.*, **161**, 1302–1314.

Early,P.W., Davis,M.M., Kaback,D.B., Davidson,N. and Hood,L. (1979) Immunoglobulin heavy chain gene organization in mice: analysis of a myeloma genomic clone containing variable and α constant regions. *Proc. Natl. Acad. Sci. USA,* **76**, 857–861.

Felsenfeld,O. (1971) *Borrelia. Strains, Vectors, Human and Animal Borreliosis.* Warren H.Green, Inc., St. Louis, MO.

Felsenfeld,O. (1976) Immunity in relapsing fever. In *The Biology of Pathogenic Spirochetes.* Johnson,R.C. (ed.), Academic Press, New York.

Fox,G.E., Stackebrandt,E., Hespell,R.B., Gibson,J., Maniloff,J., Dyer,T.A., Wolfe,R.S., Balch,W.E., Tanner,R.S., Magrum,L.J., Zablen,L.B., Blakemore,R., Gupta,R., Bonen,L., Lewis,B.J., Stahl,D.A., Luehrsen,K.R., Chen,K.N. and Woese,C.R. (1980) The phylogeny of procaryotes. *Science,* **209**, 457–463.

Francou,F. (1981) Isolation and characterization of a linear DNA molecule in the fungus *Ascobolus immersus. Mol. Gen. Genet.*, **184**, 440–444.

Hicks,J.B., Strathern,J.N. and Herskowitz,I. (1977) The cassette model of mating type interconversion. In *DNA Insertion Elements, Plasmids and Episomes.* Bukhari,A., Shapiro,J. and Adhya,S. (eds), Cold Spring Harbor Laboratory Press, New York, pp. 457–462.

Hirochika,H. and Sakaguchi,K. (1982) Analysis of linear plasmids isolated from *Streptomyces*: association of protein with the ends of the plasmid DNA. *Plasmid,* **7**, 59–65.

Holt,S.C. (1978) Anatomy and chemistry of spirochetes. *Microbiol. Rev.*, **42**, 114–160.

Hood,L., Campbell,J.H. and Elgin,S.C.R. (1975) The organization, expression and evolution of antibody genes and other multigene families. *Annu. Rev. Genet.*, **9**, 305–353.

Jacob,F. (1982) *The Possible and the Actual.* Pantheon Books, New York.

Kelly,R. (1971) Cultivation of *Borrelia hermsii. Science,* **173**, 443–444.

Klaviter,E.C. and Johnson,R.C. (1979) Isolation of the outer envelope, chemical components, and ultrastructure of *Borrelia hermsii* grown *in vitro. Acta Trop. (Basel),* **36**, 123–131.

Lalor,T.M., Kjeldgaard,M., Shimamoto,G.T., Strickler,J.E., Konigsberg,W.H. and Richards,F.F. (1984) Trypanosome variant-specific glycoproteins: a polygene family with multiple folding patterns? *Proc. Natl. Acad. Sci. USA,* **81**, 998–1002.

Meier,J.T., Simon,M.I. and Barbour,A.G. (1985) Antigenic variation is associated with DNA rearrangements in a relapsing fever borrelia. *Cell,* **41**, 403–409.

Meleney,H.E. (1928) Relapse phenomena of *Spironema recurrentis. J. Exp. Med.*, **48**, 65–82.

Plasterk,R.H.A., Simon,M.I. and Barbour,A.G. (1985) Transposition of structural genes to an expression sequence on a linear plasmid causes antigenic variation in the bacterium *Borrelia hermsii. Nature,* **318**, 257–263.

Segal,E., Billyard,E., So,M., Storzbach,S. and Meyer,T.F. (1985) Role of chromosomal rearrangement in *N. gonorrhoeae* pilus phase variation. *Cell,* **40**, 293–300.

Schoolnik,G.K., Tai,J.Y. and Gotschlich,E.C. (1982) Receptor binding and antigenic domains of gonococcal pili. In *Microbiology 1982.* Schlessinger,D. (ed.), American Society for Microbiology, Washington, DC, pp. 312–316.

Schwartz,D.C. and Cantor,C.R. (1984) Separation of yeast chromosome-sized DNAs by pulsed field gradient gel electrophoresis. *Cell,* **37**, 67–75.

Schuhardt,V.T. and Wilkerson,M. (1951) Relapse phenomena in rats infected with single spirochetes (*Borrelia recurrentis* var. turicatae). *J. Bacteriol.,* **62**, 215–219.

Stoenner,H.G., Dodd,T. and Larsen,C. (1982) Antigenic variation of *Borrelia hermsii. J. Exp. Med.,* **156**, 1297–1311.

Thompson,R.S., Burgdorfer,W., Russell,R. and Francis,B.J. (1969) Outbreak of tick-borne relapsing fever in Spokane County, Washington. *J. Am. Med. Assoc.,* **210**, 1045–1050.

Van der Ploeg,L.H.T., Cornelissen,A.W.C.A., Michels,P.A.M. and Borst,P. (1984) Chromosome rearrangements in *Trypanosoma brucei. Cell,* **39**, 213–221.

Watabe,K., Shih,M.-F., Sugina,A. and Ito,J. (1982) *In vitro* replication of bacteriophage phi29 DNA. *Proc. Natl. Acad. Sci. USA,* **79**, 5245–5248.

Watabe,K., Leusch,M. and Ito,J. (1984) Replication of bacteriophage phi 29 DNA *in vitro*: the roles of terminal protein and DNA polymerase. *Proc. Natl. Acad. Sci. USA,* **81**, 5374–5378.

CHAPTER 9

Antigenic variation in *Bordetella pertussis*

A.ROBINSON[1], C.J.DUGGLEBY[2], A.R.GORRINGE[1] and I.LIVEY[2]

[1]*Vaccine Research and Production Laboratory, and* [2]*Molecular Genetics Laboratory, PHLS, CAMR, Porton Down, Salisbury, Wilts SP4 0JG, UK*

Introduction

Bordetella pertussis, the causative agent of whooping cough, possesses a number of virulence components which allows the organism to initiate and maintain infection in the child. These components may either facilitate adhesion to the respiratory tract or produce local and systemic disease effects. The major biochemical and immunological characteristics of the principal virulence-associated components have been described in a recent review article (Robinson *et al.*, 1985) and will only be briefly summarized here.

Lymphocytosis promoting factor

LPF (otherwise known as pertussis toxin, pertussigen, histamine sensitizing factor or islets activating protein) is a major toxin associated with disease in the child. It produces a range of biological effects when injected into laboratory animals including a marked lymphocytosis, enhanced sensitivity to histamine, hyperinsulinaemia, hypoglycaemia, adjuvanticity (especially for IgE), increased susceptibility to anaphylactic shock, mitogenicity for lymphocytes and induction of experimental allergic encephalomyelitis and other autoimmune diseases. LPF is a globular protein with a molecular weight of 117 000 and is comprised of five distinct subunits. It can be isolated from cells or culture supernatants.

Heat-labile toxin

HLT, also known as dermonecrotic toxin, is cytoplasmic in origin and has pronounced dermonecrotizing activity. Although a potent toxin its role in pathogenesis is unknown.

Adenylate cyclase

This component has a molecular weight of 70 000 and is found in both extracellular and intracellular locations. It has been postulated that adenylate cyclase becomes internalized by mammalian cells causing unregulated cyclic AMP production which impairs cellular functions. This may contribute to the impaired host defences seen in *Bordetella* infections.

Haemolysin

Little is known about this component but genetic studies indicate it may be involved in virulence.

Filamentous haemagglutinin (FHA)

When viewed in the electron microscope FHA appears as fine filaments 2 nm in diameter and 40–100 nm in length. It has a high molecular weight (200 000) when examined by sodium dodecyl sulphate–polyacrylamide gel electrophoresis (SDS–PAGE) but tends to be heterogeneous. FHA has been implicated in the adhesion of *B. pertussis* to mammalian cells.

Agglutinogens

These are defined as surface antigens of *B. pertussis* which stimulate the production of antibodies that cause bacterial cell agglutination and have been used to serotype the genus *Bordetella*. There are three principal agglutinogens of *B. pertussis*. Agglutinogen 1 is common to all strains of the organism, but the presence or absence of agglutinogens 2 and 3, accounts for the four major serotypes of *B. pertussis*. An adhesion role has been implicated for agglutinogens 2 and 3, which have been demonstrated to be fimbrial proteins.

Outer membrane proteins (OMP)

Six OMP, with molecular weights of 90 000, 86 000, 82 000, 33 000, 31 000 and 30 000 have been found to be specific for the virulent form of *B. pertussis* but their role in virulence is unknown.

Recently interest in research into pathogenesis and immunology of *B. pertussis* has accelerated because of the recognized need for the production of an acellular vaccine of defined antigen composition to replace the currently used whole cell vaccine. Principal candidate antigens for inclusion in an acellular vaccine include FHA, LPF, agglutinogens, OMP and adenylate cyclase (Robinson *et al.*, 1985). Production of acellular vaccines requires growth of organisms producing high yields of the desired constituent antigens. *B. pertussis* can however give rise to changes in serotype by loss or gain of agglutinogens 2 or 3. Furthermore, antigenically deficient forms can be produced by two procedures, the phenotypic antigenic modulation and the genotypic phase variation. It is the aim of this chapter to review these antigenically altered forms of the organism and to discuss their possible relevance to the maintenance and spread of infection.

Serotype variation

There have been several reports that pure cultures of *B. pertussis* strains can display heterogeneity in the serotypes of the constituent organisms (Cameron, 1967; Stanbridge and Preston, 1974a; Bronne-Shanbury and Dolby, 1976; Preston *et al.*, 1982). Starting with single colony isolates of defined serotype from such mixed cultures, Stanbridge and Preston (1974a) detected, upon serial subculture, variants which had independently either lost or gained agglutinogens 2 or 3. Factor 1 was always retained. There is evidence that this phenomenon also occurs *in vivo*, both in experimental animal models of infection (Stanbridge and Preston, 1974b; Preston *et al.*, 1980; Robinson and Livey, unpublished observation) and in the child (Preston and Stanbridge, 1972).

Further studies are required to establish how readily interconversion between the four main serotypes (1; 1.2; 1.3; and 1.2.3) occurs and how prevalent it is among *B. pertussis* strains. It is not clear whether the serotype of an organism as defined by agglutinogens 2 and 3 is, in general, a stable characteristic (Bronne-Shanbury and Dolby, 1976) or subject to variation, perhaps at a frequency as high as 10^{-3} or 10^{-4} (Stanbridge and Preston, 1974a). With the recent demonstration that agglutinogens 2 and 3 are fimbrial proteins (Ashworth *et al.*, 1982, 1985; Irons *et al.*, 1985; Zhang *et al.*, 1985) it would be interesting to establish how serotype variation in *B. pertussis* compares with the rapid changes in fimbriae which have been reported for other bacteria (Freitag *et al.*, 1985; Hagblom *et al.*, 1985). The involvement of the minor agglutinogens, the identities of which are unknown, also needs to be elucidated.

The fimbrial nature of agglutinogens 2 and 3 makes these factors prime candidates for mediating the attachment of *B. pertussis* to the ciliated cells of the upper respiratory tract, a step presumed necessary to avoid muco-ciliary clearance. Although direct evidence is lacking on this point, monoclonal antibodies to agglutinogens 2 and 3 have been shown to inhibit binding of *B. pertussis* to Vero cells in a serotype-specific manner (Gorringe *et al.*, 1985). The development of agglutinins during the course of the infection may contribute towards the elimination of the organisms from the respiratory tract. Serotype-specific immunity induced by vaccination seems to have this effect (Stanbridge and Preston, 1974b). Serotype variation may provide the infecting organism with a limited potential to counter this facet of the host response.

Antigenic modulation

Lacey (1960) reported that changes in the growth environment could cause a reversible antigenic transition in *B. pertussis*. Growth on medium containing NaCl produced virulent organisms, the X-mode phenotype, whereas growth on medium containing $MgSO_4$ gave organisms with the avirulent C-mode phenotype. Lacey (1960) called this reversible change exhibited by *B. pertussis* antigenic modulation. He found that some ions favoured X-mode growth whereas others favoured production of C-mode organisms. An intermediate mode (I-mode) was observed under conditions intermediate between those leading to X- and C-mode growth (Lacey, 1960). However this has not been reported by other workers and it may be that populations tested were intermediate whilst the individual organisms were not.

Subsequently Pusztai and Joo (1967) observed that a similar antigenic modulation could be induced by increasing the concentration of nicotinic acid (NA) in the growth medium.

Changes observed during antigenic modulation

During antigenic modulation several components of *B. pertussis* are lost (*Table 1*). It appears that the changes during antigenic modulation caused by $MgSO_4$ and NA are very similar except that Pusztai and Joo (1967) reported a different serological response.

Table 1. Factors and attributes lost by *B. pertussis* during antigenic modulation.

Property	*References*
LPF	Parton and Wardlaw (1975); Idigbe *et al.* (1981); Robinson *et al.* (1983)
FHA	Gorringe (unpublished observations)
Agglutinogens	McPheat *et al.* (1983)
HLT	Livey *et al.* (1978); Idigbe *et al.* (1981)
X-mode specific polypeptides	Wardlaw *et al.* (1976); Idigbe *et al.* (1981); Schneider and Parker (1982); Ezzell *et al.* (1981a); Robinson *et al.* (1983); Dobrogosz *et al.* (1979)
Adenylate cyclase	Parton and Durham (1978); Hall *et al.* (1982); McPheat *et al.* (1983)
Haemolysin	Lacey (1960)
Cytochrome d629	Dobrogosz *et al.* (1979); Ezzell *et al.* (1981b)
Hydrophobicity	Robinson *et al.* (1983)
Adhesion	Burns and Freer (1982); Gorringe *et al.* (1985); Redhead (1985)

McPheat *et al.* (1983) reported that NA modulation resulted in a loss of agglutinogen factors 2 and 3, but an increase in factor 1. It has also been reported that antigenic modulation induced by NA does not produce a complete loss of HLT (Livey, 1981). The changes in OMP during $MgSO_4$ antigenic modulation of *B. pertussis* growing in a chemostat are shown in *Figure 1*.

Antigenic modulation caused by $MgSO_4$ or NA results in a change in the hydrophobicity of *B. pertussis* cells as determined by adhesion to octane (Robinson *et al.*, 1983). The X-mode cells are very hydrophobic and adhere readily to the octane droplets, whereas the C-mode cells are hydrophilic and do not adhere to octane. Furthermore, C-mode cells adhered poorly to mouse lungs (Burns and Freer, 1982), Vero cells (Gorringe *et al.*, 1985) or HeLa cells (Redhead, 1985). Monoclonal antibodies to agglutinogen 3, FHA, LPF and to an X-mode specific polypeptide partially inhibited adhesion of X-mode *B. pertussis* to Vero cells. Monoclonal antibody to agglutinogen 2 inhibited adherence of *B. pertussis* strains bearing the homologous agglutinogen (Gorringe *et al.*, 1985). This would suggest that the loss of these X-mode characteristics reduces the ability of C-mode organisms to adhere in these assays.

Environmental conditions affecting antigenic modulation

Lacey (1960) determined the modulating potential of a range of salts assessed in different ratios. X-mode growth was favoured by lithium, sodium, potassium and ammonium cations, and chloride, bromide, iodide and nitrate anions. For C-mode growth, the most effective cations were magnesium, strontium and calcium, especially in combination with sulphate, selenate, butyrate or glutarate as the anion. The I-mode was observed with X-mode cations with lactate, fumarate, malate and several other anions. Lacey also observed that C-mode growth was obtained when the Bordet–Gengou plates were incubated at 25°C. Brownlie *et al.* (1985a) examined the effect of various inorganic and organic salts on adenylate cyclase activity, histamine-sensitizing activity (HSA) and X-mode specific polypeptides. They found that $MgSO_4$, Na_2SO_4, sodium lactate, sodium succinate, sodium butyrate and sodium caprylate all caused loss of the X-mode

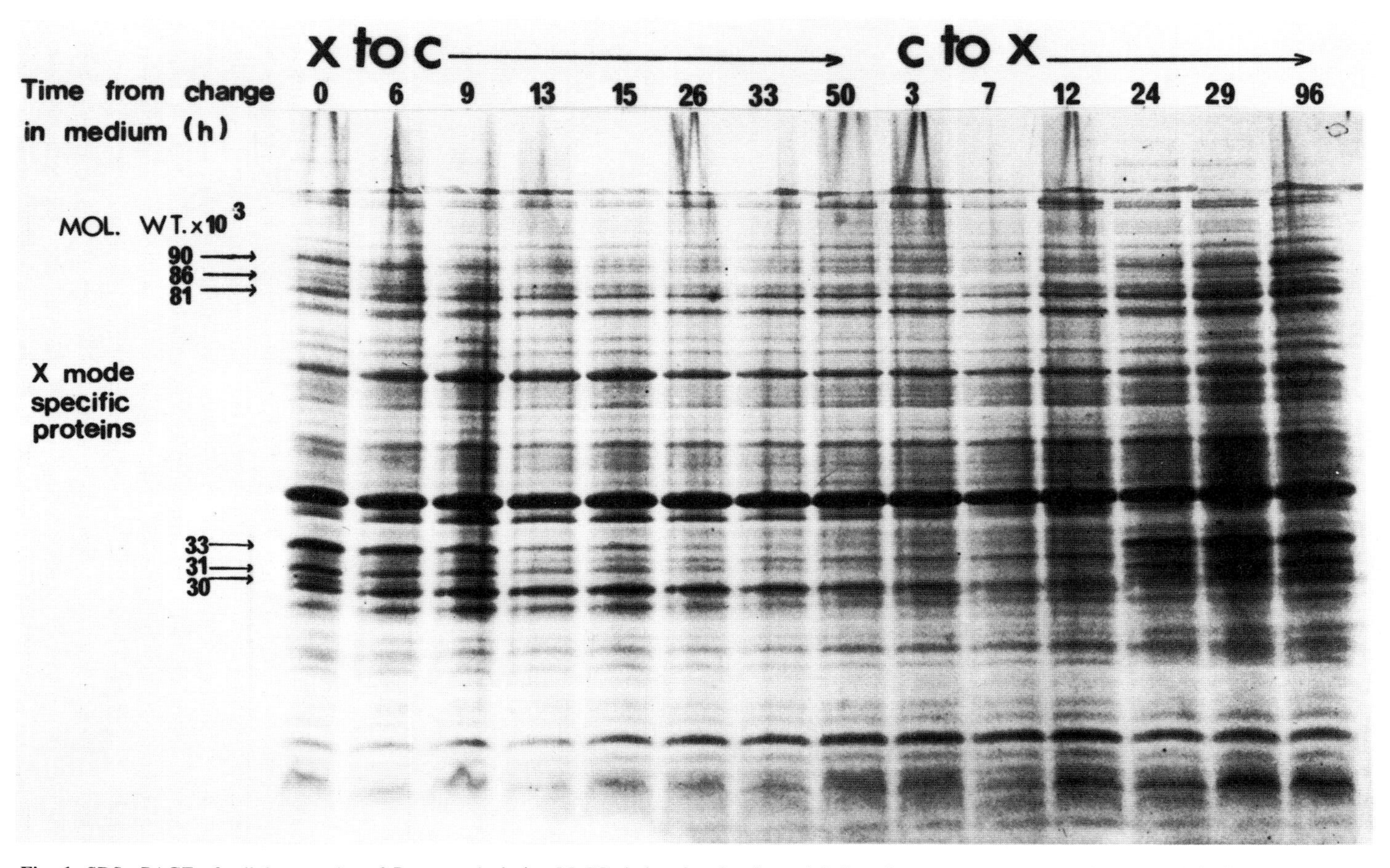

Fig. 1. SDS–PAGE of cellular proteins of *B. pertussis* during $MgSO_4$-induced antigenic modulation. At 0 h the medium was changed to C-medium and at 50 h to X-medium. Reprinted from Robinson *et al*. (1983).

characteristics, but there was no case where one property was lost independently of the others.

Pusztai and Joo (1967) reported that antigenic modulation was induced by increasing the concentration of NA in the growth medium from 0.05 to 0.5 g l^{-1}. Schneider and Parker (1982) examined several pyridines and compounds resembling pyridines for their ability to affect modulation. They found that 6-chloronicotinic acid and quinaldic acid were more effective modulating stimuli than NA on a molar basis. 2-Chloronicotinic acid and isoniazid, on the other hand, interfered with NA-induced modulation and were termed anti-modulators. Modulation cannot be induced by increasing the concentration of nicotinamide in the growth medium (Wardlaw *et al.*, 1976).

Rate of antigenic modulation

The rate of transition from X- to C-mode has been studied in both batch and continuous culture. Lacey (1960) noted that the change from X- to C-mode had occurred after 21–36 h growth on Bordet Gengou agar with a high $MgSO_4$ concentration and required 7–12 cell divisions. Idigbe *et al.* (1981) reported that strain 18334 grown in modified Hornibrook medium containing $MgSO_4$ underwent a more rapid loss of pertussis toxin-associated activities, HLT and X-mode polypeptides than could be accounted for by a simple growth-dilution effect. These workers suggested that selective destruction occurred during modulation. In these experiments there was a slow growth rate with a mean generation time of 6–7 h. The loss of adenylate cyclase activity during modulation was studied by Hall *et al.* (1982) and found to follow a curve indistinguishable from that expected if complete repression of adenylate cyclase synthesis occurred on the change from X- to C-mode medium. Brownlie *et al.* (1985a) found that the loss of adenylate cyclase, HSA and X-mode polypeptides occurred in a synchronous manner and correlated closely with the theoretical loss (assuming no synthesis) with no inactivation or destruction. They found the X- to C-mode transition took 8–10 h with a mean culture generation time of 3.2 h.

Robinson *et al.* (1983) used continuous culture techniques to study the rate of antigenic modulation. This provided a system with a constant growth rate, controlled by the dilution rate. Thus, the loss of X-mode components could be directly compared with the mean generation time and theoretical washout rate of the culture. It was found that all the X-mode characteristics assayed were lost and regained at approximately the same rate. The loss of LPF during $MgSO_4$-induced antigenic modulation, when X-mode medium in the chemostat was gradually changed to C-mode medium, was approximately equal to the washout rate at dilution rates of 0.05, 0.1 and 0.15 h^{-1} (mean generation times of 14, 7 and 4.6 h, *Figure 2*). The general consensus of opinion, with the exception of Idigbe *et al.* (1981), is that the rate of modulation induced by change to a high $MgSO_4$ medium is equivalent to the theoretical value obtained if X-mode organisms are replaced by C-mode organisms upon cell division.

For NA-induced modulation in the chemostat, C-mode-inducing conditions were instantaneously achieved by adding NA to the growth vessel (Robinson *et al.*, 1983). In this case loss of LPF and hydrophobicity was quicker than the theoretical washout rate, C-mode values of LPF content being reached after 8 h (1.14 generations) and of hydrophobicity after 12 h (1.71 generations). Loss of X-mode characteristics faster

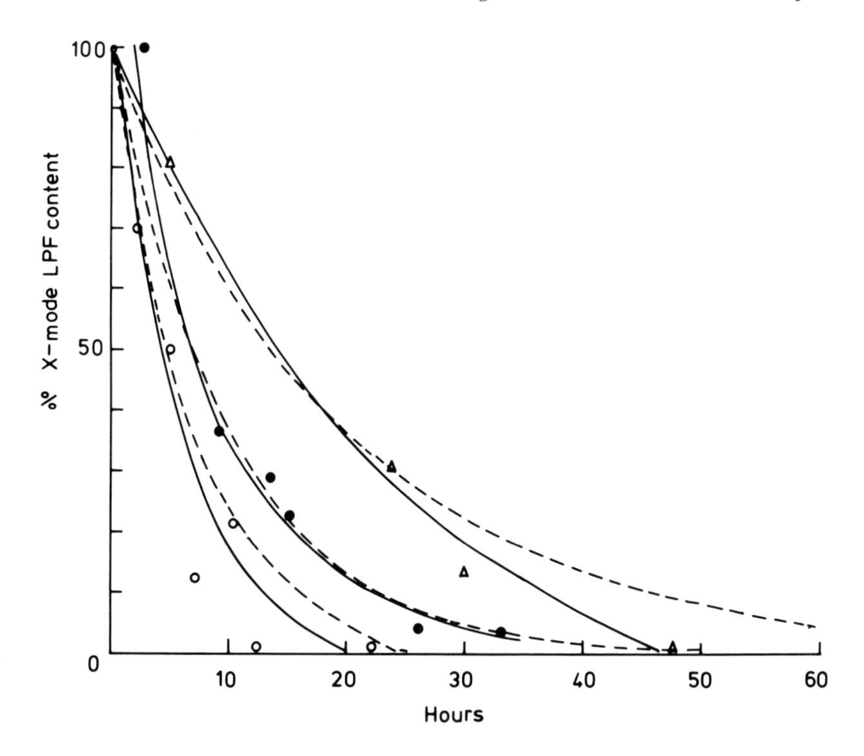

Fig. 2. The effect of dilution rate on $MgSO_4$-induced loss of cellular LPF. D=0.05 h^{-1} (△); D=0.1 h^{-1} (●); D=0.15 h^{-1} (○); dashed lines are theoretical washout rates. Reprinted from Robinson *et al.* (1983).

than the theoretical washout rate is an interesting phenomenon since it requires positive cellular inactivation or destruction of the X-mode characteristics (and conversely their active synthesis by C-mode cells at a rate faster than that predicted by the growth rate for the reverse C- to X-mode transition). Whether this phenomenon can also occur for $MgSO_4$-induced modulation (perhaps at slow growth rates) remains to be elucidated, as does the actual mechanism of active loss or acquisition of antigens.

Possible role of adenylate cyclase in antigenic modulation

Cyclic AMP, the product of adenylate cyclase, has been shown to play a regulatory role in the expression of proteins in prokaryotes (Botsford, 1981), including some virulence-associated proteins in *Escherichia coli* (Eisenstein *et al.*, 1981; Martinez-Cadena *et al.*, 1981). It has been suggested that the adenylate cyclase in *B. pertussis* may play a similar role in the control of expression of virulence factors and may be involved in the regulation of antigenic modulation (Hewlett *et al.*, 1979; Parton and Durham, 1978; Wardlaw and Parton, 1979). Brownlie *et al.* (1985a) observed that loss of adenylate cyclase occurred concomitantly with the loss of HSA and X-mode-specific polypeptides during modulation induced by different inorganic and organic salts. Further studies (Brownlie *et al.*, 1985b) indicated that adenylate cyclase does not have a causal role in the loss of virulence-related factors during modulation, but that both are influenced by the same control mechanism which produces modulation. The evidence for this is given below.

(i) The loss of adenylate cyclase activity during modulation could be accounted for by complete repression of adenylate cyclase synthesis, and there was no evidence for inhibition of adenylate cyclase activity prior to modulation.

(ii) Loss of adenylate cyclase in intact cells during modulation did not precede, but paralleled, the loss of HSA and X-mode polypeptides.

(iii) Addition of exogenous cyclic AMP or dibutyryl cyclic AMP to the growth medium had no observable effect in counteracting modulation.

Further evidence that adenylate cyclase does not act in a regulatory role in antigenic modulation stems from the observation that adenylate cyclase-negative mutants of *B. pertussis* (see below) produce normal levels of LPF, FHA and agglutinogens. If low levels of cAMP are the trigger for antigenic modulation these strains might be expected to be permanently in the C-mode.

The signal for antigenic modulation is thus not known. Furthermore, it is not known whether the control of antigen production occurs at the stage of transcription or translation. Recent advances in the study of the genetics of phase variation of *B. pertussis* have however yielded more information on the expression of virulence components of the organism.

Phase variation

Upon repeated subculture *in vitro, B. pertussis* strains may lose, in a process termed phase variation, the same virulence-associated antigens lost in antigenic modulation. However, this transition is not freely reversible in response to changes in the growth environment. The avirulent variant produced by phase variation has an increased resistance to certain antibiotics and fatty acids (Dobrogosz *et al.*, 1979; Field and Parker, 1979; Peppler and Schrumpf, 1984; Bannatyne and Cheung, 1984). The latter attribute may explain their ability to grow on media which will not support the growth of the virulent phase organism (e.g. media not supplemented with charcoal, blood, starch or cyclodextrin). The colony morphology of these phase variants is flat in contrast to the domed colonies of the virulent organism (Peppler, 1982).

There is no definitive description of the process of phase variation. Leslie and Gardner (1931) classified four antigenically distinct stages: I (as isolated) and II being toxic (to guinea-pigs) and III and IV being relatively non-toxic. Later work (Lawson, 1939; Flosdorf *et al.*, 1941; Standfast, 1951; Kasuga *et al.*, 1953, 1954) showed that the degradation process was not so simply defined. Lawson (1939) proposed that there were many more intermediates in the process of degradation, the number of which depended on the host strain and upon environmental conditions, and that the phase changes proceeded gradually, making it difficult to distinguish between the intermediate phases.

More recently, Parker (1976) proposed that only the terms 'fresh isolate', 'intermediate strain' and 'degraded strain' should be used to describe phase variation. The phase I/phase IV classification system is still however currently used to denote the extremes of the phase types, especially since not all fresh isolates should be considered as virulent and laboratory passage does not guarantee that a strain will become degraded (Weiss and Falkow, 1984).

The pattern of loss of traits on phase change

It is unclear how the transition from the virulent to avirulent form of *B. pertussis* proceeds. One view is that it is a multistep process with the various virulence factors being lost in either a random (Standfast, 1951; Parker, 1976, 1979) or an ordered way (Leslie and Gardner, 1931; Goldman *et al.*, 1984). Alternatively traits may be lost and regained in a single step event (Weiss and Falkow, 1984).

Parker (1976, 1979) proposed that phase variation is the result of a non-ordered accumulation of multiple point mutations which select for variants deficient in virulence properties but with an enhanced capacity to grow *in vitro*. This is consistent with the observation of Standfast (1951) that fresh isolates of *B. pertussis* differed in the rate and order in which they lost virulence-associated functions upon prolonged subculture. However, it is difficult to know how the parameters he studied (immunogenicity, haemagglutination, mouse virulence/toxicity, ability to grow on blood agar) relate to the virulence factors now recognized and how to reconcile the data with that of more recent observations.

To further investigate the phenomenon of phase variation, Peppler (1982) selected variants of *B. pertussis* which had acquired the ability to grow on Stainer and Scholte agar, a medium which would not normally support *B. pertussis* growth. On transfer to Bordet–Gengou agar these variants were non-haemolytic and had a flat colony morphology, in contrast to the domed haemolytic colony type which typifies the growth of virulent *B. pertussis* on this medium. Further characterization revealed that the variants lacked FHA, LPF, adenylate cyclase, agglutinogens and certain OMP specific for the virulent form of *B. pertussis*. These avirulent organisms were present in wild-type populations at a frequency between 5×10^{-5} and 5×10^{-6}. Goldman *et al.* (1984) found that variants of strains Tohama and 165 which grew on media lacking any detoxification agents occurred at a frequency of $10^{-6}-10^{-7}$. Upon screening 250 of these isolates for the presence of haemolysin, LPF and FHA, four classes of variants were obtained. One type of variant, which had apparently been noted by others (Field and Parker, 1979; Dobrogosz *et al.*, 1979), had all three virulence factors (7–11%). The other variants lacked either haemolysin alone (17%), both haemolysin and LPF (5–11%), or all three traits (65%). This latter and most prevalent type of variant would seem to be identical to the fully avirulent form of *B. pertussis* identified by Peppler (1982). Goldman *et al.* (1984) proposed that isolates with some or all virulence factors are intermediates in a multistep process which eventually gives rise to the fully avirulent organism. Since all three traits studied occurred in only four of eight possible combinations it was concluded that the process was non-random. However, the existence of organisms of intermediate phenotype does not necessarily mean they are intermediates in the process of phase variation and further experimentation is required to confirm or refute this model of phase variation.

Weiss and Falkow (1984) used increased tolerance to erythromycin as a virulence phase marker to select *B. pertussis* colonies which lacked haemolysin, LPF and FHA. The proportion of these avirulent forms in the virulent phase population was estimated to be between 10^{-3} and 10^{-6}, depending on the strain. This change was reversible since spontaneous phase variants, of the same phenotype as those selected using erythromycin, were occasionally found to revert to the haemolytic form. Such isolates had also regained FHA, LPF and sensitivity to erythromycin. Since revertants could only be screened for by the reappearance of haemolytic activity, and not directly selected, it was not possible to measure the frequency of reversion. Moreover, reversion seemed to be influenced by environmental factors since certain batches of media seemed to favour the selection of the virulent organisms. This may explain why others who have identified similar avirulent variants to those described by Weiss and Falkow (e.g. Peppler, 1982; Goldman *et al.*, 1984) have not noted reversion. Weiss and Falkow (1984) were able to demonstrate several cycles of variation starting with an avirulent variant, indicating that it was a readily reversible single-step event. The single-step nature of this event was further confirmed by the observation that a single insertion of transposon Tn*5* at a particular site in the *B. pertussis* chromosome could effect the concomitant loss of not only LPF, FHA and haemolysin but also adenylate cyclase and HLT.

In conclusion, it appears that the loss of virulence factors from *B. pertussis* can occur in a single-step, reversible process. Whether there are other ways in which fully avirulent organisms can arise remains to be established. The loss of virulence factors *en bloc* does not appear to be concomitant with the acquisition of the ability to grow on nutrient agar, a feature of classical phase IV isolates. The fully avirulent Tn*5* mutant of Weiss and Falkow (1984), BP 347, does not grow on nutrient agar (McPheat and Brownlie, personal communication). Similarly, Peppler's avirulent variants selected by growth on Stainer and Scholte agar did not grow on nutrient agar, although *B. pertussis* capable of growth on nutrient agar could be selected from the latter variants at a rate of between 10^{-7} and 10^{-8} per cell per generation (Peppler and Schrumpf, 1984). Such organisms had further increased resistance to oleic acid and various antibiotics but could not be directly selected from the wild-type organism. Thus, growth on nutrient agar seems to be an additional attribute acquired in a process distinct from the loss of virulence characteristics.

A model for phase variation

The observation that there can be a concomitant loss of several characteristics might suggest that the loss of genetic information results from the curing of a plasmid or prophage encoding the virulence factors. However, a correlation has not been shown between plasmid content and virulence in *B. pertussis*. Furthermore the reversibility of the observed phase change (Weiss and Falkow, 1984) is not compatible with this mechanism of phase variation. As an alternative one might consider a model in which the virulence-associated genes are arranged in one operon with their expression being under the control of a single promoter. Such an arrangement is not consistent with the results of studies using transposon mutagenesis with Tn*5*. Insertion of Tn*5* into a gene on such a polycistronic operon would effect the loss of not only the virulence factor encoded by the target gene but also those factors encoded by genes downstream from the insertion site, since Tn*5* has been shown to cause polar mutations (Berg *et al.*, 1980). Yet most virulence mutants obtained by Tn*5* mutagenesis lack only a single virulence factor. Weiss *et al.* (1983) found that haemolysin, FHA and LPF can be lost independently by Tn*5* mutagenesis, an observation that we have confirmed and find can be extended to include agglutinogen 3 (*Table 2*). Furthermore, Tn*5* insertions producing haemolysin, FHA and LPF mutants have been mapped within different restriction frag-

Table 2. Tn*5* derivatives of *B. pertussis* strain Wellcome 28 deficient in virulence-associated factors[a].

Factor(s) lost/reduced	*Number of mutants*
Haemolysin	1 (partial loss)
Haemolysin + adenylate cyclase	1
FHA	4
LPF	2
Agglutinogen 3	6

[a]Approximately 1100 isolates were screened for haemolysin, adenylate cyclase, FHA, LPF, agglutinogens 1, 2 and 3 and the virulent phase-specific 33-K OMP.

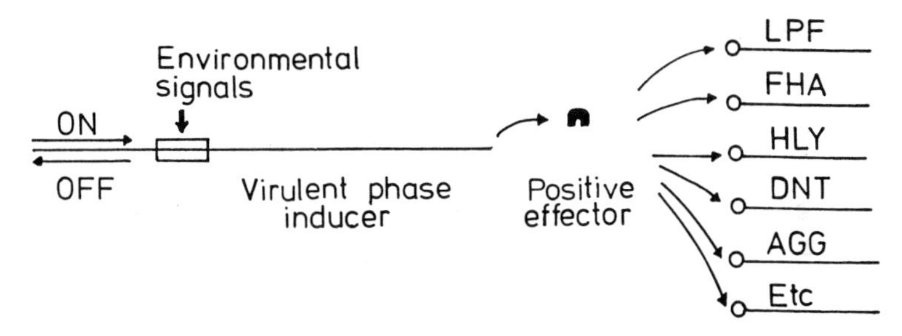

Fig. 3. A model for phase variation and antigenic modulation. Production of a *trans*-acting positive effector switches on the genes encoding the virulence factors. The effector could be the protein encoded by the virulent phase-inducer gene or a molecule generated as a result of this protein's activity (cf. regulatory role of cAMP in control of transcription). The effector itself could be subject to regulation in response to changes in environmental conditions (antigenic modulation) and by a mechanism which is largely independent of environmental influences (phase variation). It is postulated that an invertible region of DNA may control phase variation by switching between an 'on' and 'off' position. Redrawn from Weiss and Falkow (1984).

ments in the *B. pertussis* chromosome indicating that they might not be closely associated (Weiss *et al.*, 1983). Only the haemolysin and adenylate cyclase have been found to be associated in that Tn*5* mutants selected as being non-haemolytic often, but not always, lack or have reduced levels of adenylate cyclase (Weiss *et al.*, 1983; *Table 2*). This latter information is consistent with the haemolysin and adenylate cyclase genes being on one operon with the adenylate cyclase structural gene being proximal to the promoter.

The model proposed by Weiss and Falkow (1984) is based on the premise that there exists a *trans*-acting gene product which is required for the expression of the virulence-associated genes. Support for the existence of a positive effector stems from the observation that a single insertion of Tn*5* into one particular region of the *B. pertussis* chromosome can result in the loss of many (if not all) of the virulence factors. This model would allow for coordinate expression of the virulence-associated genes but does not require that they are in the same transcription unit (*Figure 3*). *B. pertussis* in the avirulent phase would not produce the proposed *trans*-acting gene product. This model would also provide an explanation for the failure to obtain expression of *B. pertussis* virulence factors when fragments of *B. pertussis* chromosomal DNA are cloned into *E. coli* (C.J. Duggleby, unpublished observation; Weiss and Falkow, 1983). Expression would require the cloning of not only the structural gene for a virulence factor but also the gene for the positive effector, an unlikely event if the latter gene is not closely associated with the virulence genes on the chromosome.

To account for the reversibility of the phase transition it is envisaged that the virulent phase inducer is itself subject to regulation. One possibility is that its expression is controlled by an invertible region of DNA which can switch between an 'on' and 'off' position, in a manner analogous to the genetic switches of other bacteria and bacteriophage (see Saunders, Chapter 5). At present there is only limited evidence for the existence of an invertible element in *B. pertussis* with homology to other invertible elements (e.g. *Gin, Hin, Cin, Pin*; Foxall, Drasar and Duggleby, unpublished observations) and none for DNA rearrangements. An alternative proposal is that phase variation can be effected (and presumably reversed) by spontaneous mutation in a controlling region (Lax, 1985).

One additional merit of the model proposed by Weiss and Falkow (1984) is that it is also a model for antigenic modulation. Thus, it is proposed that environmental signals may be able to modulate the expression of the postulated *trans*-acting gene product and thus affect the expression of the virulence factors. By screening for mutants unable to modulate, it may be possible to analyse how different growth conditions are able to trigger the process of antigenic modulation.

Conclusions

A precise role for antigenic modulation or phase variation in the maintenance or spread of infection of *B. pertussis* has not been satisfactorily described. Antigenic changes in other bacteria have been considered to be a means of evading the host's developing immune response. With phase variation of *B. pertussis* the loss of all virulence characteristics would seem to be too drastic a method to evade an immune response since avirulent non-infective organisms are produced. *B. pertussis* may adapt to the host's immune response by antigenic changes that are restricted to changes in serotype of the infecting organism (i.e. loss or gain of agglutinogens 2 or 3). It is generally accepted that pertussis vaccines should contain both agglutinogens 2 and 3 to provide full protection against all major serotypes of *B. pertussis*.

If phase variation and antigenic modulation are not a means of evading the immune response, they may promote transmission of the disease. In this respect the readily reversible antigenic modulation may be the more important characteristic. Virulence components such as FHA, agglutinogens and X-mode-specific OMP are required for the organism to adhere to the respiratory tract of the child and resist clearance mechanisms such as mucociliary flow and coughing. It is interesting to speculate that as disease progresses, local tissue damage and possibly the enzymic action of LPF may give rise to local environments that favour the formation of C-mode organisms. It has been demonstrated in a number of laboratories that C-mode organisms adhere less well to mammalian cells than X-mode cells and thus would be expelled by the child, facilitating transmission of the disease. Such a hypothesis depends on two assumptions which are testable by experimental observations. (i) Tissue damage, etc. gives rise to local environmental conditions that favour the production of C-mode organisms; (ii) C-mode organisms are capable of initiating infection in animal models and reverting to the virulent X-mode form. Should these observations be confirmed experimentally then it would be reasonable to assume that antigenic modulation of *B. pertussis* is a means of evading the immune response by ensuring spread of the disease to a new host. Phase variation may simply be due to laboratory-acquired mutations in the gene(s) of *B. pertussis* organisms which confer upon them the ability to undergo antigenic modulation.

References

Ashworth,L.A.E., Irons,L.I. and Dowsett,A.B. (1982) Antigenic relationship between serotype-specific agglutinogen and fimbriae of *Bordetella pertussis. Infect. Immun.*, **37**, 1278–1281.

Ashworth,L.A.E., Dowsett,A.B., Irons,L.I. and Robinson,A. (1985) The location of surface antigens of *Bordetella pertussis* by immuno-electron microscopy. In *Proceedings of the Fourth International Symposium on Pertussis, Developments in Biological Standardization. Vol. 61,* Manclark,C.R. and Hennessen,W. (eds.) S. Karger, Basel, pp. 143–151.

Bannatyne,R.M. and Cheung,R. (1984) Antibiotic resistance of degraded strains of *Bordetella pertussis. Antimicrob. Agents Chemother.*, **25**, 537–538.

Berg,D.E., Weiss,A. and Crossland,L. (1980) Polarity of Tn5 insertion mutations in *Escherichia coli. J. Bacteriol.*, **142**, 439–446.

Botsford,J.L. (1981) Cyclic nucleotides in procaryotes. *Microbiol. Rev.*, **45**, 620–642.

Bronne-Shanbury,C.J. and Dolby,J.M. (1976) The stability of serotypes of *Bordetella pertussis* with particular reference to serotype 1,2,3,4. *J. Hyg.*, **76**, 277–286.

Brownlie,R.M., Parton,R. and Coote,J.G. (1985a) The effect of growth conditions on adenylate cyclase activity and virulence-related properties of *Bordetella pertussis. J. Gen. Microbiol.*, **131**, 17–25.

Brownlie,R.M., Coote,J.G. and Parton,R. (1985b) Adenylate cyclase activity during phenotypic variation of *Bordetella pertussis. J. Gen. Microbiol.*, **131**, 27–38.

Burns,K.A. and Freer,J.H. (1982) Mouse lung adhesion assay for *Bordetella pertussis. FEMS Microbiol. Lett.*, **13**, 271–274.

Cameron,J. (1967) Variation in *Bordetella pertussis. J. Pathol. Bacteriol.*, **94**, 367–374.

Dobrogosz,W.J., Ezzell,J.W., Kloos,W.E. and Manclark,C.R. (1979) Physiology of *Bordetella pertussis.* In *International Symposium on Pertussis.* Manclark,C.R. and Hill,J.C. (eds), US DHEW Publication No. (NIH) 79-1830, Washington, DC, pp. 86–93.

Eisenstein,B.I., Beachey,E.H. and Solomon,S.S. (1981) Divergent effects of cyclic adenosine 3′,5′-monophosphate on formation of type I fimbriae in different K-12 strains of *Escherichia coli. J. Bacteriol.*, **145**, 620–623.

Ezzell,J.W., Dobrogosz,W.J., Kloos,W.E. and Manclark,C.R. (1981a) Phase-shift markers in *Bordetella*: alterations in envelope proteins. *J. Infect. Dis.*, **143**, 562–569.

Ezzell,J.W., Dobrogosz,W.J., Kloos,W.E. and Manclark,C.R. (1981b) Phase-shift markers in the genus *Bordetella*: loss of cytochrome d629 in phase IV variants. *Microbios,* **31**, 171–182.

Field,L.H. and Parker,C.D. (1979) Differences observed between fresh isolates of *Bordetella pertussis* and their laboratory passaged derivatives. In *International Symposium on Pertussis.* Manclark,C.R. and Hill,J.C. (eds), US DHEW Publication No. (NIH) 79-1830, Washington DC, pp. 124–132.

Flosdorf,E.W., Dozois,T.F. and Kimball,A.C. (1941) Studies with *H. pertussis* V. Agglutinogenic relationships of the phases. *J. Bacteriol.*, **41**, 457–471.

Freitag,C.S., Abraham,J.M., Clements,J.R. and Eisenstein,B.I. (1985) Genetic analysis of the phase variation control of expression of Type I fimbriae in *Escherichia coli. J. Bacteriol.*, **162**, 668–675.

Goldman,S., Hanski,E. and Fish,F. (1984) Spontaneous phase variation in *Bordetella pertussis* is a multistep non-random process. *EMBO J.*, **3**, 1353–1356.

Gorringe,A.R., Ashworth,L.A.E., Irons,L.I. and Robinson,A. (1985) Effect of monoclonal antibodies on the adherence of *Bordetella pertussis* to Vero cells. *FEMS Microbiol. Lett.*, **26**, 5–9.

Hagblom,P., Segal.E., Billyard,E. and So,M. (1985) Intragenic recombination leads to pilus antigenic variation in *Neisseria gonorrhoeae. Nature,* **315**, 156–158.

Hall,G.W., Dobrogosz,W.J., Ezzell,J.W., Kloos,W.E. and Manclark,C.R. (1982) Repression of adenylate cyclase in the genus *Bordetella. Microbios,* **33**, 45–52.

Hewlett,E.L., Underhill,L.H., Vargo,S.A., Wolff,J. and Manclark,C.R. (1979) *Bordetella pertussis* adenylate cyclase: regulation of activity and its loss in degraded strains. In *International Symposium on Pertussis.* Manclark,C.R. and Hill,J.C. (eds), US DHEW Publication No. (NIH) 79-1830, Washington, DC, pp. 81–85.

Idigbe,E.O., Parton,R. and Wardlaw,A.C. (1981) Rapidity of antigenic modulation of *Bordetella pertussis* in modified Hornibrook medium. *J. Med. Microbiol.*, **14**, 409–418.

Irons,L.I., Ashworth,L.A.E. and Robinson,A. (1984) Release and purification of fimbriae from *Bordetella pertussis*. In *Proceedings of the Fourth International Symposium on Pertussis, Developments in Biological Standardization. Vol. 61*. Manclark,C.R. and Hennessen,W. (eds), S.Karger, Basel, pp. 153–163.

Kasuga,T., Nakase,Y., Ukishima,K. and Takatsu,K. (1953) Studies on *Haemophilus pertussis*. I. Antigen structure of *H. pertussis* and its phases. *Kitasato Arch. Exp. Med.*, **26**, 121–134.

Kasuga,T., Nakase,Y., Ukishima,K. and Takatsu,K. (1954) Studies on *Haemophilus pertussis*. III. Some properties of each phase of *H. pertussis*. *Kitasato Arch. Exp. Med.*, **27**, 37–48.

Lacey,B.W. (1960) Antigenic modulation of *Bordetella pertussis*. *J. Hyg.*, **58**, 57–93.

Lawson,G.McL. (1939) Immunity studies in pertussis. *Am. J. Hyg.*, **29**, (Section B), 119–131.

Lax,A.J. (1985) Is phase variation in *Bordetella* caused by mutation and selection? *J. Gen. Microbiol.*, **131**, 913–917.

Leslie,P.H. and Gardner,A.D. (1931) The phases of *Haemophilus pertussis*. *J. Hyg.*, **31**, 423–434.

Livey,I. (1981) Ph.D. Thesis, Glasgow University.

Livey,I., Parton,R. and Wardlaw,A.C. (1978) Loss of heat-labile toxin from *Bordetella pertussis* grown in modified Hornibrook medium. *FEMS Microbiol. Lett.*, **3**, 203–205.

McPheat,W.L., Wardlaw,A.C. and Novotny,P. (1983) Modulation of *Bordetella pertussis* by nicotinic acid. *Infect. Immun.*, **41**, 516–522.

Martinez-Cadena,M.G., Guzman-Verduzco,L.M., Stieglitz,H. and Kupersztoch-Portnoy,Y.M. (1981) Catabolite repression of *Escherichia coli* heat-stable enterotoxin activity. *J. Bacteriol.*, **145**, 722–728.

Parker,C. (1976) Role of the genetics and physiology of *Bordetella pertussis* in the production of vaccine and the study of host-parasite relationships in pertussis. *Adv. Appl. Microbiol.*, **20**, 27–42.

Parker,C.D. (1979) The genetics and physiology of *Bordetella pertussis*. In *International Symposium on Pertussis*. Manclark,C.R. and Hill,J.C. (eds), US DHEW Publication No. (NIH) 79-1830, Washington, DC, pp. 65–69.

Parton,R. and Durham,J.P. (1978) Loss of adenylate cyclase activity in variants of *Bordetella pertussis*. *FEMS Microbiol. Lett.*, **4**, 287–289.

Parton,R. and Wardlaw,A.C. (1975) Cell-envelope proteins of *Bordetella pertussis*. *J. Med. Microbiol.*, **8**, 47–57.

Peppler,M.S. (1982) Isolation and characterization of isogenic pairs of domed hemolytic and flat nonhemolytic colony types of *Bordetella pertussis*. *Infect. Immun.*, **35**, 840–851.

Peppler,M.S. and Schrumpf,M.E. (1984) Isolation and characterization of *Bordetella pertussis* phenotype variants capable of growing on nutrient agar: comparison with phases III and IV. *Infect. Immun.*, **43**, 217–223.

Preston,N.W. and Stanbridge,T.N. (1972) Efficacy of pertussis vaccines: a brighter horizon. *Br. Med. J.*, **3**, 448–451.

Preston,N.W., Timewell,R.M. and Carter,E.J. (1980) Experimental pertussis infection in the rabbit: similarities with infection in primates. *J. Infect.*, **2**, 227–235.

Preston,N.W., Surapatana,N. and Carter,E.J. (1982) A reappraisal of serotype factors 4, 5 and 6 of *Bordetella pertussis*. *J. Hyg.*, **88**, 39–46.

Pusztai,Z. and Joo,I. (1967) Influence of nicotinic acid on the antigenic structure of *Bordetella pertussis*. *Ann. Immunol. Hung.*, **10**, 63–67.

Redhead,K. (1985) An assay of *Bordetella pertussis* adhesion to tissue culture cells. *J. Med. Microbiol.*, **19**, 99–108.

Robinson,A., Gorringe,A.R., Irons,L.I. and Keevil,C.W. (1983) Antigenic modulation of *Bordetella pertussis* in continuous culture. *FEMS Microbiol. Lett.*, **19**, 105–109.

Robinson,A., Irons,L.I. and Ashworth,L.A.E. (1985) Pertussis vaccine: present status and future prospects. *Vaccine*, **3**, 11–22.

Schneider,D.R. and Parker,C.D. (1982) Effect of pyridines on phenotypic properties of *Bordetella pertussis*. *Infect. Immun.*, **38**, 548–553.

Stanbridge,T.N. and Preston,N.W. (1974a) Variation of serotype in strains of *Bordetella pertussis*. *J. Hyg.*, **73**, 305–310.

Stanbridge,T.M. and Preston,N.W. (1974b) Experimental pertussis infection in the marmoset: type specificity of active immunity. *J. Hyg.*, **72**, 213–228.

Standfast,A.F.B. (1951) The phase I of *Haemophilus pertussis*. *J. Gen. Microbiol.*, **5**, 531–545.

Wardlaw,A.C. and Parton,R. (1979) Changes in envelope proteins and correlation with biological activities in *B. pertussis*. In *International Symposium on Pertussis*. Manclark,C.R. and Hill,J.C. (eds), US DHEW Publication No. (NIH) 79-1830, Washington, DC, pp. 94–98.

Wardlaw,A.C., Parton,R. and Hooker,M.J. (1976) Loss of protective antigen, histamine-sensitizing factor and envelope polypeptides in cultural variants of *Bordetella pertussis. J. Med. Microbiol.*, **9**, 89–100.

Weiss,A.A. and Falkow,S. (1983) The use of molecular techniques to study microbial determinants of pathogenicity. *Phil. Trans. R. Soc. Ser. B,* **303**, 219–225.

Weiss,A.A. and Falkow,S. (1984) Genetic analysis of phase change in *Bordetella pertussis. Infect. Immun.*, **43**, 263–269.

Weiss,A.A., Hewlett,E.L., Myers,G.A. and Falkow,S. (1983) Tn5-induced mutations affecting virulence factors of *Bordetella pertussis. Infect. Immun .,* **42**, 33–41.

Zhang,J.H., Cowell,J.L., Steven,A.C., Carter,P.H., McGrath,P.P. and Manclark,C.R. (1985) Purification and characterization of fimbriae isolated from *Bordetella pertussis. Infect. Immun.,* **48**, 422–427.

INDEX